IT
REALLY
DID
HAPPEN

The changes in farming since the 1940's

To MY COUSIN GWEN

A. H. McElwaine

Alan McElwaine

ISBN: 1-905451-26-1

A CIP catalogue for this book is available from the National Library.

Printed in Ireland

This book was published in cooperation with
Choice Publishing & Book Services Ltd, Ireland
Tel: 041 9841551 Email: info@choicepublishing.ie
www.choicepublishing.ie

Dedication

I would like to dedicate this book, firstly to my Uncle, Roy Graham, who started it all (see top photo – front cover).

Secondly to my wife Verna, who has backed me and supported me right from the start. For the past two years she has put up with my papers, notes and drafts scattered all around the house. She also forgave me if I suddenly put on the light in the middle of the night, to scribble down some thought that had just occurred to me.

Finally, to my three daughters, who have inspired me and helped me with the typing and research. They also acted as "sounding boards" to bounce ideas off.

In conclusion, let me thank them for five wonderful grandchildren (so far!!). Here's hoping that at least one of them will pick up the "Farming Bug".

Acknowledgements

I am indebted to many people who provided assistance in the course of researching and writing this book.

My first thank you goes to Eugene Markey – County Museum, Ballyjamesduff, Co Cavan. If it had not been for the encouragement given to me by Eugene, the book would probably have stopped at four or five chapters and most likely would never have been published.

My next thank you goes to Seàn Ò Roideáin, who at the time was General Manager of Bailieborough Development Association (B.D.A.). He offered to have his clerical staff transfer my notes and scribbles onto Floppy Disc. Shortly after this he left the B.D.A. and was replaced by the present Manager, Jim Maguire. Thankfully, he agreed to continue with the arrangement.

The following girls from B.D.A. have done trojan work; Ciara Mc Keon, Elizabeth Fleming, Mary Mc Donald and Leona Mc Donnell. I do not know how they managed to decipher my handwriting.

Another person who was most helpful was David Cockburn B.Ag., Teagasc Office, Bailieborough. He kept me correct concerning A.C.O.T., Teagasc etc.

Next I come to Brian Reynolds, Photographer, Barrack Street, Bailieborough. The photographs and artwork were left in his capable hands.

For all the above, and also to my wife and family – I say a very big "Thank You."

BAILIEBOROUGH DEVELOPMENT ASSOCIATION LTD

Barrack Street
Bailieborough
Co Cavan
Telephone No: 042-9694716 Fax No 042 9694717
Website : www.bailieborough.com
Email : bda@bailieborough.com
Vat No. IE 8236009F
Charity Reg. No. CHY11546

Bailieborough Development Association is an organisation, set up and managed by members of the Bailieborough Community. Our aim is to bring services to the town and people of Bailieborough where a need has been identified.

Our projects to date include – Out of School Childcare Service, Training and Development Centre, Recycling and Amenity Centre, and a Custom Built Business Centre offering quality rentable office space to business.

We are constantly seeking ways to assist in the growth and development of Bailieborough town and its community.

Preface

It has been said that everyone has at least one book in them, the problem lies in getting this book out onto paper. My story has been ready to come out for some time for the following reasons. Firstly, it all began about ten or twelve years ago. Whenever I mentioned to my daughters (3) about something that happened back in the fifties the reaction normally was "Dad, that never happened", or "that could never happen". Well it really did.... hence my title.

Secondly, if I waited another ten or fifteen years to write it there would be very few people around who actually milked cows by hand, made hay with a hay fork and hand rake, or tackled a horse and cart!

Thirdly, there are people who think that everything that is in use today was "invented" in the last ten years or so. Not true... see chapter entitled Nothing New.

Lastly, if the building and development around Bailieborough continues at it's present level for two or three years there will no longer be farmland around the outskirts, as I knew it.

This is not a history lesson, not even a geography lesson, just a mix 'em gather 'em of memories from someone who was there. I know there will be those amongst my readers who will not agree with everything I have written, or my interpretation of the facts. On the other hand, there will be those who will say that I left out a lot of important material.

All I can say to both sides is "write your own book!"

For the benefit of those amongst my readers who have never heard of Bailieborough, it is situated in East County Cavan, about 30 miles from the Border with Northern Ireland.

In 2006 it has a population of approximately 2000 – 2500. Back in 1940 this would have been 1000 – 1200. At that time, like any small market town, it was dependant on the farming community in its hinterland.

Glossary

Word	Meaning	Chapter
Alpacas	Like a Llama	Alternative Enterprise.
Bottans	Bundle of Hay	Graham's Yard
Bonhams	A Young Pig	Fair Day, Grahams Yard, Pigs
Britchen	Part of Harness for working horses	Horses
Boxty	Like Potato Bread	Tillage 2
Caseog	First Litter Sow	My Farm
Clarendo	Flaked Maize	Animal Feeds
Clogs	Wooden Soled Shoe	Fair Day
Capon	A neutered cockrel	Poultry
Chats	Small potatoes	Tillage 2
Eiderdown	Old fashioned Duvet	Poultry
Epicures	A variety of potato	Tillage 2
E.T.	(No, not the little brown fellow)	Breeding Livestock
F.Y.M.	Farmyard Manure	Grahams Yard
Fiddle	Not a violin	Tillage 1
Galloshes	Rubber over shoes	Preparing for Fair Day
Haggard	Small field	Tillage 1
Hames	Part of harness for working horse	Horses
Hand Shakes	Small cocks of hay	Haymaking
Jennet	Offspring of Donkey Mare & Pony Stallion	Horses

Glossary

Word	Meaning	Chapter
Landrace (Finnish)	Breed of sheep	Sheep
Landrace (Swedish)	Breed of Pig	Sheep
Laps	Bundle of wet hay	Haymaking
Mule	Offspring of Pony Mare & Donkey Stallion	Horses
Martingale	Part of Harness for riding horse	Horses
Mangold	Like a turnip	Tillage 1
Odeama	Disease in pigs	My farm
Old Grannies	Stooks of Corn	Tillage 1
Pike	(Not the fish) Stack of Corn	Tillage 1
Rasp	A coarse nail file	Forge
Sharpes Express	Variety of potato	Tillage 2
Sprigs	Small nails for repairing shoes	Preparing for Fair Day
Split Kip	Leather for working boots	Preparing for Fair Day
Tray of Bread	Six loaves in batch	Saturday deliveries
Tangler	A pig dealer	Fair Day
Tumbling Paddy	A wheel less Hay Rake	Hay Making
Whang	Leather boot lace	Preparing for Fair Day
Water Glass	Egg preservative	Poultry
Zero Grazing	Feeding Cows indoors	Nothing new

Table of Contents

Chapter 1

Bailieborough Town

People complain about the traffic on the Main Street of Bailieborough, especially at the weekends. I would like to compare the present day with 50 years ago. Back to the traffic jams – now they are caused by cars, delivery vans etc. In 1950 they were mostly at their peak early in the morning and again in the late afternoon. Believe it or not, they were caused mostly by cows being brought in and out for milking – yes cows! In those days practically every house in town or at least on the Main Street kept some form of livestock in their back yards. When I speak of livestock I am not thinking of dogs, cats or even pigeons. I am thinking more of cows, horses, pigs and hens. It would take too long to name every household in town, I will just focus on the main characters.

If we start at Lower Main Street, what is now, the "Corner Court", was then, "Birney's Pub". They milked cows. They also had a lodger Willie Roundtree, a cattle dealer, whose lorry was normally parked around the corner, along the Courthouse wall, - more about that later.

We move on now, to Willie Bell. I never knew of him to have his own animals but he was an agent for Donnellys Bacon Company. The farmers used to deliver their fat pigs and on the appointed day every week, the noise was horrific as all these pigs were loaded onto a lorry to be transported to the bacon factory. However, Willie Bell's main connection with the farming community was as an Animal Feed Compounder. As well as supplying pig and cattle feed to the Bailieborough area, his lorries were on the road every day, delivering feed all over counties Cavan, Monaghan, Meath and Louth. Speaking of lorries, Willie had a problem, namely a narrow gateway! His fleet of 10 Ton Rigid Lorries would bring a load of barley from the docks in Dublin, but when they arrived in Bailieborough, these lorries could not fit through the gateway. The load had to be off loaded on the street, onto smaller pickup trucks, which then reversed into the Mill in the yard. When you consider that we are talking about 16 stone (100kg) bags, with no means of mechanical handling, you can appreciate the task. This task had to be repeated in reverse when the compounded feed was coming out again.

Our next port of call is Matt Traynors, now known as "The Attic". He kept pigs at the bottom of his garden. A large part of their diet was raw eggs! Traynor's operated a mobile shop around the country. Often when they supplied tea, sugar etc to the farmer's wife, she would pay with eggs. These eggs were sold in Dublin. In fact they had a contract to supply the canteen in Trinity College Dublin. Before they could be sent to Dublin, the eggs had to be tested to see that they were fresh. To do this, all eggs had to be passed on a conveyor belt in front of a strong electric bulb. Anything that looked suspect was broken and thrown into a bucket – hence the pigs. Matt also bought and sold turkeys at Christmas. There was always great rivalry between him and Jim & Joe Mitchell (now J&L Stores). At the start of the Turkey season they both had a problem – if they did not pay enough they would not get many turkeys but if they paid too much they would get turkeys but they would not be able to sell them on. One year Jim Mitchell decided he would get smart – he got a young boy from the country to go down to Matt Traynor to say "Mammy wants to know how much you are paying for turkeys". Unfortunately the youngster got a bit excited and asked "What are you paying for turkeys". Matt suspecting that he might be a messenger from Mitchell's said "Tell them I am paying money".

Next door we had Cecil McKinley – Saddler and Harness Maker. Farmers and Horsemen for miles around came to get their harnesses made and repaired.
He also repaired school bags, ladies handbags etc. Cecil had land in different places around the town. It was common then to see him on a Wednesday afternoon, when the shops were closed, driving his cattle down Main Street on their way from Rakeevan to Drumbannon.

Three doors up we had another interesting yard, belonging to Paddy Kangley, father of Pat. He normally kept his cattle out on the farm but I remember that occasionally, in winter, he would house some of them in the Main Street. However, the reason Paddy Kangley's yard, and the adjoining yard of Willie Coote's, father of Bert, stick in my memory is on Fair Day in March, April and May it became a hub of Equine activity. This was when the two yards were rented out as Stallion quarters. From early morning until late in the evening, the yards were packed with mares, coming to visit the stallions. My pals and I spent many an afternoon watching the activities, much to the annoyance of said Willie Coote as he thought it was not proper for youngsters to witness this. This operation also took place in Tom Murtaghs Yard.

Rose Sullivan also kept cows, pigs and hens. The hens were allowed run around the yard. Sometimes the gate to the Main Street would be left open and the hens would get out onto the street, much to the delight of the dogs of the town.

The next place I can remember animal activity was in Shaffrey's Hotel (now The Bilberry). They kept cows and pigs. Next door, (now The Credit Union) we had Harry McCartneys Pub and Grocery. Harry milked a number of cows, mostly pedigree Dairy Shorthorns. He also kept pedigree Hereford cattle for beef. He was a good judge of "horse flesh" and kept both horse and pony mares for breeding. He also kept a black and white jack ass "King Jacko". Harry was a great Agricultural Show man and was a regular exhibitor at the RDS in Dublin. Harry also had a corn mill where he rolled or ground oats and barley for the farmers.

At the top of Market Square we had Mary Geelans Pub. She kept sows. On Market Square we had Lynchs Butcher (Hughie the boot) now John Ed Sheanon. As well as having their slaughterhouse there, they also kept animals in the yard. Across the street Mrs Pat Brady kept pigs and broiler chickens. We also had a butcher's stall on the street.

The Bailie Hotel was then Grahams Hotel and grocery. They had a very varied farm yard with, cows, calves, sows, pork pigs, hens and a working pony called "Tiny". Tiny, a small black pony was later replaced by a brown and white skewbald called Bunny named after an RTE radio character called Bunny O'Hanlon. Roy Graham was one of the first in the Bailieborough area to keep the then new fangled, black and white cows called Friesian. He was also the first in the area to keep laying hens in cages. People then and still today say this was cruel and unnatural, but the hens didn't seem to mind, they layed every day sometimes twice and if for some reason they did not lay today they would lay 3 eggs tomorrow. Roy's wife Lily used to churn the milk and serve fresh homemade butter to the Hotel guests. This was made into little balls or "pats" with a pair of wooden bats called Scotch hands. They also put a jug of fresh buttermilk on the dinner table in the hotel everyday. More about Grahams later.

An End over End churn

Two doors down, we had Tom Murtagh, father of Peter. They milked cows and I seem to remember they fed pigs as well. Tom also had a corn mill where he ground oats and barley. As I said earlier he also "let out", his yard as stallion quarters on fair day. Another clear memory of Tom Murtagh was that he bought Blackberries. In August – September the kids of the area would arrive in with Blackberries in sweet cans or in buckets. It might take a whole day to fill a bucket but when they were weighted at Murtagh's you might get a half crown (15c). When weighed the Blackberries were emptied into, what seemed to us, huge wooden barrels. We never quite knew what Tom did with them. Some said they went to either Fruitfield or Lambs jam factory, others said they went to make dye for cloth.

Next door to Tom Murtagh, we had Paddy Argue the butcher (Blacks today). As a child I would be sent up on Saturday morning to get the meat for the weekend. Mr Argue had a very annoying habit of leaving children standing while he served adults. I often had to stand about for an hour or more. When I eventually got served he would say, through his nose "run along now sonny and don't say it was here you were kept". Another saying of his, if someone complained about there being bones in the meat they were buying was, "If you buy land you got to buy stones and if you buy beef you got to buy bones".

Two doors down from the butcher we had T.D. Vance, Solicitor (now Mel Kilraine). T.D. did not keep any animals here, but at his residence on the Virginia road he kept Jersey cows and Hunting horses. His son Cecil later became Chairman of the now defunct Bailieborough Co-op. Cecil and his wife were both keen followers of "the hunt" and all things equestrian. Their son Thomas later became an international show jumper.

Paddy Farrell the butcher, kept cows and pigs. His brother Owen next door also milked cows. Both of them had land out beside the town lake. One of them, I think it was Paddy had a very well trained dog that could go for the cows on his own, managing to get the gate open, drive the cows out and bring them along Church street up Main Street and into their yard on his own. Going back after milking someone had to go with him because he could not close the gate on his own. .

T.R.Smyth (McDonagh Shoes) milked his cows out in the country and then sold the milk around the town, either from a donkey and cart or later from a little Austin van. A few years later T.R. was the first in Bailieborough to be an agent for bottled milk, which he delivered door to door in the same little Austin van. The milkman delivered 4 bottles and collected the 4 empty bottles from the day before. Some people would forget to leave out their empties; this created a problem for the dairy, as it was expensive to have to keep replacing the glass bottles. The Milk Bottlers Association later ran an advert on new fangled television "Your milk comes in a bottle, we can't bring round the cow, please wash them and return them as most of you do now."

An earlier generation, an uncle of T.R.'s did milk his cows in town. In order to give air to his cows, small ventilation holes were cut in the wall above the cows' head. These holes came through to McElwaine's yard next door. My grandmother often told how, if, she found herself with unexpected guests and no milk or cream for their tea, she would place a jug in one of these holes and then knock the wall to draw the attention of Mrs Smyth. Very quickly the jug was lifted in, filled and put back in the hole, to be retrieved by my grandmother.

Where SuperValu is now, we had Tom Tierney. He fed pigs and gradually built up his stock until at the end, he was feeding 25-30 bacon pigs at a time. On one occasion Tom was bringing a trailer load of pigs to McCarren's Bacon Factory in Cavan. Somewhere along the way, part of the tailboard of his trailer fell off. Before he got to Cavan one of the pigs jumped out. Needless to say he had quite a job getting it reloaded to continue his journey.

Where Planet Earth and the Chinese Restaurant are now, we had John James McCleary. He milked cows, fed pigs, hens and ducks. The space between Planet Earth on one side and Mrs Patsy Bells house was open ground. It was a lot lower than Main Street and it had a little stream running through it. Mrs. McClearys ducks used to have a "right" time swimming in this stream.

In the fifties, the largest milking herd in the town belonged to James Kangley. He lived in the corner house between Ann Street and the Back Road. At that time he milked 10 or 12 cows and supplied milk to a large part of town. My mother got her milk there for years. However in the early fifties people began to get health conscious and became aware of the connection between bovine T.B and human T.B. The state scheme for bovine T.B. testing had not started at this time. However, Roy Graham had started to have his cows tested privately, and so my mother transferred her custom there.

Where J&L Stores is now, was originally the premises of the Mitchell Bros, Jim & Joe. They bought and sold eggs and all types of poultry. They were both keen gunmen and the spoils of their sport, namely rabbits and pheasants were often on sale.

Turning into Barrack Street I remember the late Dick Murray, father of Brian kept cows, hens and possibly pigs. Tom O'Dowd's father also lived in Barrack Street and milked cows. He owned the field between the Kells/Virginia Roads where the new houses are currently been built.

Chapter 2

Outskirts of Town

If we continue along the Virginia Road we come to Trinity Presbyterian Church. In the forties and on in to the fifties there was a large galvanised shed at the back of the church, one half of this contained 8 or 10 stables or stalls where the parishioners could tie their horses when they drove to church on a Sunday morning, the other half was large enough to back their traps into. I know these premises were still used until the late fifties but the only people that I remember using it on a regular basis was James Lyttle, father of Wilkie, and Mr and Mrs Forbes who lived out the Kingscourt Road.

Just past the church on the opposite side of the road we had the convent. They milked cows. I don't mean to say that the nuns milked the cows, they had a "servant boy" to do it for them. The cows were grazed on the fields now used by Pat McIntyre. Where "The Mews" houses are built now was where the nuns had their hens, ducks and sometimes a sow.

On the Kingscourt Road, just at the edge of town we had the late Michael Trainor, Bexcourt. He milked a couple of cows. They had to be walked up Thomas street, through Market Square up past Masonic Hall and down Curkish lane or the Bottle lane as it was called then, to get to his land in Rakeevan (all built up now). The reason Curkish Lane was called the Bottle Lane was because the triangular piece of ground between Curkish Lane and the Mountain Road was at that time the town dump, known as the Bottles. Sometimes bottles, tin cans etc. would roll from the dump down onto the lane, hence the name Bottle Lane.

About 200 yards over on the Bottle Lane there used to be a stream running across the lane. In summer there would be a steady trickle, but in winter there could be quite a flood. While the stream was an inconvenience to anyone walking along the lane, it had its usefulness as well. When Mick

Trainor was walking his cows in and out for milking they could get a drink here. Roy Graham also had land beside this stream and his cows drank from here as well. People travelling the lane with a horse/pony and cart also used it as a watering hole.

Speaking of watering holes, on the Kingscourt Road opposite Mick Trainors there was a horse trough along the side of the road. This was built in such a way that no matter how low the level of water in the little stream went, there would always be enough water in the Horse trough to allow any horse or cow passing to get a drink. This trough was at the entrance to McCartneys field.

McCartneys field was where the circus used to visit. Then the horse trough was used by elephants, camels, llamas etc. In the fifties the circus arrived by horse power, the tents, caravans and animal cages were all pulled by teams of horses. Before the late show was even over the "crew" started to dismantle the tent and pack it away. There would be great excitement in town, sometimes as early as 6 or 7 am, as we heard the clip clop of the horse's hooves on the tar as the circus arrived.

Chapter 3

Milking

I have talked a lot about cows being walked in and out for milking. This was not to a modern milking parlour with tiles around the walls and a milking machine to do the job. Instead it was an ordinary cow byre with whitewashed walls, and the cows were milked by hand. The two main pieces of equipment necessary were a stool and a bucket. In this locality the stool was a "four legged affair", although I believe that the proper milking stool was round and had "three legs". I have also heard of a "one legged stool" which the milker strapped around his/her waist, and when they got up to move from one cow to the next the stool automatically came with them. Well as I said, you need a stool and a bucket, you sit on the stool with the bucket between your knees, and away you go, milking two spins or teats at a time. A very fast milker would probably need to change buckets half way through, as fast milking means more froth.

Most of the people in town who kept cows used the milk for their own household and the remainder would go to the calves and pigs. Some sold it across the counter in their shops. This was raw or unpasteurised. Any surplus milk would be sent to the creamery. This was the cause of another traffic jam, milk from farmers out in the country was transported through town on its way to the creamery, collecting and bringing the town milk as well. These "carters" as they were called, transported the milk by horse and cart, no refrigerated bulk tanks just 9, 10 and 12 gallon galvanised cans. Each carter might have milk from 12 -15 farmers on his cart. Each farmer had a creamery number that was painted on his cans so as to keep each farmer's milk separate when they got to the creamery. On their way to the creamery with a full load these carters went "Straight thro", but on the return journey they stopped to fill their shopping list. Their clients along the way in the morning would ask them to procure various items for them. It might be a loaf of bread, a bucket for feeding calves, a pair of

Wellington boots or 2lb of wire nails. As very few farmers at that time had cars and, as a result, seldom got to town they depended on the postman or the milkman to do their shopping. I was speaking to a "carter" recently who operated with a tractor and trailer in 60-70's. He told me how on occasions he had to milk the cows if he arrived and the milk was not ready. He told me that he was paid 2-3 old pence (about ½ c) per gallon for collecting and delivering the milk to the creamery. This was deducted out of the farmer's cheques at the end of the month and paid to the carter. He told me about a widow woman he used to collect and draw for, who lived fairly near to town. She objected that she should have to pay as much "cartage" as a farmer living 3 or 4 miles out. She negotiated a price for herself, which she agreed to pay direct to the carter rather than have it deducted at the creamery. The only problem was that it sometimes went 3-4 months before she paid.

In the fifties there were two bothers, the Barnes who carted milk with a horse and cart. They later progressed to a little grey Ferguson 20. Whenever one of their milk clients was getting married, or even a son or daughter of a client, the Barnes always bought a wedding present. The thing was they always bought the same present – an E.PN.S. Teapot

Some farmers, if they had more milk than was normal at the time, would deliver their own milk. In the fifties, this would have been with a horse and cart. By the sixties, these men had got tractors and so delivered their milk by tractor and transport box. Also in the late fifties or early sixties, the carters progressed to tractor and trailer. By the late seventies, all milk had to be transported in a refrigerated bulk tank.

I knew of one farmer who insisted in delivering his own milk. He maintained that, with the amount of milk he had, he could not afford to pay the 2 or 3 pence per gallon being charged by the carters. What he failed to take into account was the fact that he had the price of diesel, plus wear and tear to his tractor. He also bought the daily paper every morning as he passed through the town. Some days, he would even have a "pint" with his mates. He then stopped in town talking for a further half an hour or more. Considering all of the above, the round trip of approximately 10 miles (16km) from his farm to the creamery and back again, could take anything up to 2 ½ hours, so I could never figure how it paid him to deliver his own milk!

I should have said that it was only suppliers living in fairly close proximity to the creamery that had the option of delivering their own milk or have it collected by a carter with tractor and trailer. Suppliers from further out were collected by lorry. This was originally in 10 – 12 gallon cans. Then the lorries changed to refrigerated, articulated trucks.

In 2004, the creameries decided that individual farmers coming with their own milk were causing too much of a traffic hold-up around the milk in take point, so self-delivery was outlawed. Then in 2005, the carters drawing with tractor and trailer also got the "chop". Now all milk has to be collected in refrigerated, articulated trucks.

This caused a lot of hardship to some farmers, especially those living in narrow lanes. Some had to give up milk production, others had to redesign their farmyards to make room for these big lorries to turn. A small number of producers who did not wish to go out of production, but at the same time could not accommodate the big lorries, had to arrange to bring their milk out to the end of their lanes.

Chapter 4

Grahams Yard

I said I would return to Graham's yard. This was where I spent a lot of my youth, and where I got the foundation stone of my love for and knowledge of farming. It was here I milked my first cow, learned how to tackle and un-tackle a horse and cart, watched my first sow farrowing (having pigs) etc. Roy normally milked four cows but occasionally this went up to six. As I said earlier he was one of the first in this locality to keep Friesians (see chapter on Dairy cows). The calves from these cows were reared, and some extra calves were also bought in. The cows were well trained and normally attended to the call of nature on request. It was important that they did not do this in the yard as hotel guests sometimes parked their cars in the yard and they did not wish to walk into "you know what". Similarly we did not want the cow's visiting card in front of hotel door. Now for the solution!! If we gave each cow a sharp slap with the palm of our hand, just above the head of her tail, before opening her chain, she would oblige within seconds (normally). If not we then had to run her out and well up the street so fast that she could not concentrate on two things at a time!! Roy kept hens in the then new fangled "Battery Cages". They were in rows of individual cages. Each hen had her own feed supply and water supply. There was a number 1 to 100 on each cage and when the eggs were collected each day a record was kept of which hens laid and which did not. As I said earlier if a hen did not lay today she would normally lay two eggs or maybe three tomorrow. If she failed to lay for a full week, you first of all checked to see that she was not sick, and if not, she was soon despatched as chicken soup.

Also on the poultry side Roy kept "table birds". He had a long loft divided into about 6 or 7 sections. Each section was large enough to accommodate about 25-30 adult birds. You started off with 25-30 day old chicks in section 1. When they were about 2 weeks old you then put new day olds in section 2, 2 weeks later section 3 etc. till you got to the end of the loft. By the time the last section were 2 weeks old the first section were about ready for slaughter and so the whole process started again. This way he always had fresh chicken on the menu in the hotel.

When the day old chicks were placed in the pen a small area was divided off using stiff cardboard, to keep the little chicks near the heat. In the early days this heat was provided by "an oil burner hoover". As the chicks started to grow and change from "Fluff to Feathers" the heat was gradually reduced. By the early sixties the oil burner was replaced with an electric infra red bulb. This was much easier to operate and it overcame the smell of oil about the chicken house.

Next in the yard we come to the pigs. There were normally 3 or 4 sows kept, which meant that there was always a mixture of pigs about the place, from "suckers" a few days old right up to bacon weight. In fact where the "outdoor smoking area" is presently situated in the Bailie Hotel is where the bacon pens were.

The main reason for keeping pigs was to use up the waste food from the hotel. Of course they needed other food as well as "slops". Most days there had to be a large pot of "spuds" boiled for the pigs. Roy had an unusual way of boiling these spuds – a sawdust boiler. Sawdust was always freely available from one of the local sawmills. To make a boiler you needed a 30 gallon steel barrel. You cut one end out of the barrel completely and just cut a round hole out of the other end. Next you placed the handle of a shovel into the barrel pushing it out through the hole at the bottom. Holding the handle to keep it straight you then filled the barrel with saw dust, packing it tightly around the shovel handle. When the barrel was full, it was lifted up onto a brick or flat stone, one on either side, so that it was raised off the ground. You then removed the shovel handle slowly. If the sawdust was well enough packed you finished up with a clear hole or tunnel right the way from the top of the barrel thro to the bottom. Next you placed two metal bars across the top of the barrel, got a large saucepan, enough to hold

40 - 50 kgs of raw potatoes, place potatoes in pot, cover with water and then lift and place on top of the metal bars on barrel. You then got a small empty tin (say a baked bean tin), place a piece of rag in the tin and add some paraffin oil, placed the tin under the hole in barrel and lit the rag with a match. The flame from the rag would be sucked up into the tunnel in the sawdust setting fire to the sawdust. The sawdust slowly burnt and by the time the sawdust had all burnt your potatoes were cooked. It was advisable to leave the "pot" to cool before lifting it down, otherwise a nasty scald could be the outcome. When it had cooled, you lifted it down, drained off the water, and then mashed the potatoes using a large stick or "beatle". The potatoes were then ready for feeding to the pigs.

Later on this concept of a sawdust boiler was put to another use. During the seventies Grahams Hotel registered with the newly formed Bord Failte. One of the conditions that they insisted upon was that the Hotel must provide hot and cold running water to every bedroom. The sawdust boiler was quickly converted. The principal remained the same, except that instead of a pot of spuds on top of the barrel you had a specially constructed copper cylinder full of water which was plumbed into a normal copper cylinder, upstairs in the hotel. The water from the copper cylinder was circulated around the bedrooms. The sawdust barrel was filled and lit first thing in the morning and last thing at night and there was lots of hot water available. There was a problem sometimes if a guest required hot water in the middle of the afternoon, otherwise the system worked very well for a number of years.

During the winter months when the cows were indoors all the time they were fed on hay. Silage had not got as far as Bailieborough in the fifties. The hay was normally brought from the hayshed morning and evening as required. However, on Saturday evening you brought round enough to do till Monday morning. In order to avoid waste the hay was made into "Bottans". This was done by lifting an armful of loose hay, enough for one feed for a cow, you then took a few more strands of hay twisted this into a sort of hay rope and wrapped it around your armful of hay and tied. You then brought these bottans and stacked them in the corner of the cow byre and at feeding time you threw one bottan in front of every cow. Needless to say that once baling of hay started there was no longer any need to make bottans. In winter the cows were also fed turnips, swedes and mangolds. These were put through a "pulper" and came out in long pieces, rather like chipped potatoes. These were fed raw.

Turnip pulper

Grahams always had a working pony. This was used for various tasks, from delivering groceries from the shop to outlying customers, to drawing in hay from the fields, to drawing out farmyard manure (F.Y.M.). In the fifties the pony was a brown and white (skewbald) called Bunny. There were two carts, a "spring cart" which was light and as the name would suggest had springs attached to make for a smoother drive. It was used for delivering groceries or even for "pleasure driving". The other cart, known as a "heavy cart" or stiff cart was stronger and used for tasks where the load to be carried was heavier and speed was not important.

Chapter 5

Saturday Deliveries

The shop deliveries took place on different days of the week depending on the route involved. The road I remember most was the Mountain run as it always took place on Saturday when I was off school.

Let me explain about these deliveries. As I said earlier, in the late forties, early fifties, very few people out in the country had cars or tractors, and so got to town very seldom. They would get small bits and pieces brought out by the postman or milk carter, but the shop in which they dealt delivered their large weekly order regularly.

Back to Saturday morning. Willie Preston, long dead, and I spend till about eleven o'clock loading up the cart. Most got a regular order every week, but occasionally they would send a message in with the postman to include something extra in their order. All orders were arranged in large wicker baskets. There was no need to put names on the baskets, as they were loaded onto the cart in house order. We delivered to the same houses every week and in the same order, so the first basket into the cart was for the last house on the run and so on till the last basket loaded would be the first delivered. We put a white cloth over each basket as it was loaded to keep the contents clean. I should remind my readers that at that time there was very little wrapping used on groceries. When the full load was on the cart we then put on a waterproof cover to cover everything against the weather. The only two things on the cart not under this cover were Willie and I. This was fine in summertime but it was a different matter in winter. I should say that to an enthusiastic 10 year old the thrill of driving Bunny up the Mountain Road was much more important than the possibility that we might get wet.

To take a look into these baskets they all had one thing in common – mainly bread. At that time the pan load was just coming in, but it was unsliced and unwrapped. The most popular bread was "Batch bread". Today some bakeries bake and sell a so- called "batch loaf", it is very different to the batch bread of the fifties. The name in those days came from the fact that they were baked in batches of six. When this bread was delivered into the grocery shop by the baker, it was still in batches of six, known as a tray of bread. The shopkeeper then pulled a loaf off the tray, or two loaves which ever their customers required. Back to our baskets, most contained at least one tray, others 2 trays (12 loaves). When you consider that they did not have electricity and therefore no fridges or freezers you can imagine what the last of the bread was like before we arrived the following Saturday. The basket also contained tea, sugar etc. Let me say that the tea being delivered was not Barry's, Bewleys or Liptons, it was Grahams. You see, the grocer bought his tea in large tea chests (containing about 50 kgs tea) which they weighed out into ¼ lb bag and ½ lb bags. A lot of houses still have "tea chests" lying around. They may hold sheets and blankets or they may be in the coal-house for holding "slack" or the farmer may store tools and implements in one. Next time you see one you will know that it did once hold tea!!

The same story applied to sugar, it came in 50kg jute sacks and was then weighed out. In those far off days, biscuits, sweets, prunes, raisins etc all came loose and had to be weighed out.

To get back to the Mountain Road, one of our last calls was to a Mrs Smith (I think she was grandmother of T.M.Smith, Auctioneer). When we arrived she always had tea and a boiled egg ready for us. At first I could not figure out how she knew the exact time of our arrival. Then I discovered that it was "Bunny" who was giving us away. When we started on the road from our last call Mrs Smith would hear the clip-clop of the pony's shoes on the tar and would then know when to expect us.

Chapter 6

Animal Feeds

Another way Bailieborough has changed is in the number of shops selling animal feed. In the 50's there were 14, today two. At that time meal was delivered to the shops in 50kg jute sacks. Most shopkeepers, in good weather would display the brand of feed they stocked out on the footpath in front of their shop. The bags would be opened and the top rolled down. Now a certain lady from the Market Square area would sometimes let her sows out onto the street to find their own feed. They might start with a feed of dairy meal at the first shop they came to and when driven away from there move on to the next shop front where it might be layers mash. Needless to say neither the lady nor her sows were very popular. As I said meal was delivered in 50kg bags. Some shopkeepers would weigh this out in smaller amounts to suit their Customers' needs. Towards the end of the 50's when most people could afford to buy the full bag, the practice of weighing out began to cease. However, there was one lady, again from the upper end of town, kept one sow and liked to buy her meal fresh. She would be the first customer into Roy Grahams shop on a Monday morning for a ¼stone (approx3lbs) of sow and weaner meal to feed her sow. She would then be the last customer in the evening with the same order. The only time she bought more than one feed at a time was Saturday evening when she was forced to buy ¾ stone. At the time she was the only customer Roy had who required weighing out. He tried to explain this to her, requesting that she should buy the full bag. "No thanks Mr Graham, I prefer to get it fresh each feed". She was then told that once the bag was opened for her, and as nobody else got any out of it, at ¼ stone twice daily it was the same bag of meal at the middle of the month as she got at the first of the month. Thinking that the problem might be money he offered to give her the full bag and she could pay for it on the drip (at least that way he was saved the problem of weighing out). Even this offer was turned down. It took a long time to change this lady.

In the fifties there were a lot more brands of feed. The following brands have gone forever. Willie Bell (already mentioned), Ranks/Blue Cross, Boland's, Bestock, Spicers of Navan, Newman's of Athboy, Farringtons of Rathcoffely, Kavanagh's of Maynooth to name but a few. I should point out however that at least two new mills have started up in the same period. These are Corbyrock Mills on the outskirts of Monaghan Town and McCabes of Canningstown.

The types of meal were also different. Cubes or nuts as we know them today had not been invented. Most calves then were fed on a gruel or porridge made with Linseed Meal. Another popular feed at that time was "yellow meal" also known as "Indian meal" now known as Maize Meal (still available but very difficult to procure.) What we know today as "Flaked Maize" or "Cooked Maize" was known then as Clarendo.

In the forties, fifties and early sixties meal was delivered in 50kg jute sacks. These sacks had the name, address and company logo stencilled on the outside. If these sacks were returned to the original compounder they were washed and reused. If they came back to a different compounder the first name would have the word "cancelled" stencilled across it. After cleaning, the second compounder then stencilled his name etc on the other side. Next time it might come back to a third compounder. He turned the sack inside out put his name on and away it went again.

In the early sixties the paper sack appeared. At first these were still 50 kg. As they were only used once they were more hygienic (less risk of spreading disease from one farm to another). However they created a major problem for the farmer, he had to dispose of them. At the beginning of the sixties another major change took place. This was the introduction of "Bulk Delivery". In effect, this meant that if a farmer ordered 5 tonne of dairy feed, instead of getting 100x50 kg bags, he got one loose delivery blown by an air pump into his feed loft. This meant that the compounder did not have to buy 100 sacks nor had he to pay staff to fill these sacks. Instead he just blew it onto an enclosed lorry.

When the lorry got to the farm another blower transferred it from lorry to loft. This meant that he could offer his feed to the farmer at a better price.

Soon free standing bulk bins began to appear all over the country. They were available in 5 tonne to 20 tonne capacity. Some of these bins were sub-divided so that the farmer could have Dairy feed in one side of the bin and Calf feed in the other side.

The transition to bulk delivery did not go without a hitch. I worked at that time for a National Compounder based in Dublin. The Workers Union always had a strangle hold on Management. I should have said that another advantage to the compounder was the fact that it only required one man on the lorry to operate the blower, as compared to two if the load had to be taken off by hand.

Back to our Dublin compounder. The Union feared that the new system would eventually lead to redundancies. The only way they would agree to bulk delivery, initially, was with a system called "Rip and Tip". This involved filling the feed into sacks as normal, build said sacks on lorry in normal way, and send two men on the lorry in traditional way. When the lorry arrived on the farm these two men then proceeded to carry the sacks to the back of the lorry, rip the bag open and tip the contents into the blower, hence the name "Rip and Tip". The company had no alternative but to agree to this, if they wished to hold on to their bigger customers. However, they decided shortly afterwards to close down the Dublin mill and relocated in the country, where the Unions did not have as severe a grip.

The delivery of meal in bulk was initially confined to the larger pig and poultry farmers. However, dairy farmers soon began to get the message. There was a problem however, the compounders would only deliver in bulk if the farmer was in a position to order a minimum of 4-5 tonnes at a time. Needless to say this was too much for the average dairy or beef farmer in Co Cavan at that time.

This problem was solved a few years later with the arrival of the "One Tonne Bulk Bag". For this system to operate, the local "Agricultural Merchant" either Co-operative or Private, agreed to erect a number of 20 tonne Bulk Bins on his premises. The feed compounder then delivered and blew into these bins. By having a number of bins the merchant could carry a selection of different types of feed, from different compounders.

When the farmer required feed he could order 1 tonne "Brand X" dairy nuts and 1 tonne "Brand Y" Calf Nuts. These were then filled into bulk bags and weighed. These bags could then be delivered by the merchant's lorry or the farmer could collect them with his tractor and trailer.

A further refinement of the "Bulk Delivery" system was the use of barrels, both steel and plastic and various other metal containers. With these containers the merchant had to weigh them when empty, then fill/weigh again, and subtract the first weight (tare) to arrive at the amount of the feed supplied. With the use of these smaller containers "bulk" was available to everyone regardless to the size of their enterprise.

At the beginning of 2002 the "powers that be" introduced what most lay people considered a silly rule: 1 tonne bulk bags could only be used once. In other words if a farmer ordered 1 tonne bulk for the first time he had to buy a new bulk bag. Under the old system he continued to reuse that bag, virtually 'till it fell to pieces. Now under the new rules he was obliged to buy a new bag every time.

This new ruling was as a follow-up to the "Foot and Mouth" scare we had in 2001. The Department Officials were very conscious of cross contamination from one farm to another. The thinking in this case was that if a farmer brought his "Bulk Bag" in to be refilled and while on the merchant's premises it came in contact with "Bulk Bags" from another farm one could get cross contamination between the bags, which in turn could lead to cross contamination between farms.

The farmers could accept the logic of this ruling. However, what annoyed them was the fact there was no similar restriction placed on barrels etc. Barrels and other containers could be reused for years on end, regardless of the state of hygiene or lack of it.

I said at the start of this discussion on "Bulk" that there was a cost saving to the farmer, compared to buying his feed in 25 kg bags. However if the farmer has to buy a new "bulk bag" every time he orders feed, this saving is largely wiped out. The outcome of this is that while bulk bags are still in use more farmers are switching to various types of steel or plastic containers.

Chapter 7

My Farm

I myself kept pigs and hens at the back of our house on Main Street. My father or grandfather never kept any animals but a previous owner of the property obviously had because the shed I converted to a pig house had previously been used as a cow byre. The first poultry that I had was a clocking hen and a clutch of small chicks (as far as I can remember there were 7 chicks), which I received as a present from the late Mrs Robert Gilmore, Lear. The male chicks eventually ended in the roasting dish while the females went on to become laying hens. I kept laying hens for a number of years after that. My first venture into pigs consisted of 4 suckers bought from a Tangler (see chapter on Fair day), at £4 – 2- 6 each, (about €5.15 today's money). One died from oedema after about a week. This was a very common disease of pigs at that time. The remaining three were sold 3 months later for I think £8 each. After paying for meal they just about broke even. In the house I converted I could feed 10 or 12 at a time. I discovered that the old town sewer ran through our garden about 3 feet away from the door of my "pig sty". This was an old sewer made with flat stones rather than red clay pipe, which was in use at that time or red plastic as used today. As I said this sewer was no longer in use but it did run into the present system. I opened into it and I used to sweep the effluent from the pigs into it. Once a week, usually on Wednesday (½ day in town) I would get a hosepipe and wash everything away. Anyway, the county council got to know that I fed pigs and they came to inspect the "premises". When they discovered that I was washing everything into the old sewer I really got "down the banks". I was told that the effluent from one pig was the equivalent of 8 humans so if I was feeding 12 pigs it was the same as increasing the town population by approx 100. I was ordered to cease production immediately. I did stop for a while but when the "dust settled" I went back into production for a couple of years.

I later changed from feeding Pork and Bacon Pigs to keeping young sows or caseogs as they were known. This involved buying young females or slips as they were called, bring them to the Boar when they were ready, feed them for a further 3 months and then sell them before they farrowed (had their litter). The nearest Boar Pig or hog was kept by the late Phildy Dunne, father of Terry. To get my caseog out there I would tie a rope to one of her back legs and walk her out. I usually left her over night and if the visit was successful I would walk her home again next morning. Sometimes it was necessary to leave her there for 3 or 4 days. The reason for the rope was to prevent the caseog from running away.

At that time, the yard until recently occupied by Austin Clarke, was a garden belonging to the late Mr and Mrs Crean. Mrs Crean ran a drapery shop, where Sandi Modes is now. Her husband Charlie had gone to England to get work and only came home once or twice a year. The result was that their vegetable garden had become a wilderness. I got permission from Mrs Crean to run my sows in this garden. As there was only a hedge between her garden and ours, it was an easy matter to run the sows out every morning and home again at night. I got special leather harnesses made to enable me to tether them in Creans, otherwise they would keep running back to our garden and if they dug up my mothers flowers I would not have been very popular.

Chapter 8

Blacksmiths Forge

In the same yard as Harry McCartney's we had Phil Sheridan's Blacksmith's Forge. All day long you could hear the tap tap of his hammer on metal as he made horse and pony shoes. When the horses etc were brought in to get shoes or to be "shod", as it was known as, the metal shoe was firstly heated in the fire. This fire had a bellows or hand operated fan attached which would make the fire extremely hot. When the shoe was taken out it would be "red hot". It was placed against the hoof until it scorched the hoof and left the shape of itself on the hoof. The shoe was then removed and left to cool. The blacksmith then lifted the horses hoof, held it between his knees and with a special curved knife, pared the hoof to the shape of the shoe. The shoe was then nailed to the hoof (you know the old rhyme "For the want of a nail the shoe was lost, for the want of a shoe the horse was lost etc. etc.). When the shoe was fully fitted, any rough edges of hoof were removed using a RASP or file.

There were two other blacksmiths in town at the time, Crossan's – opposite the Masonic Hall and Ned Kavanagh in Adelaide Road also known as Australia Street. The latter could only shoe ponies or donkeys, because there was a very narrow entry into his forge and large horses could not get in.

Also at that time we had Corries Forge about two miles out of town. As well as the usual shoeing of horses, they also made Horse Drawn Ploughs, Drill Ploughs, Hay Rakes and Gates (ornamental as well as field).They later advanced to make machinery for Tractors, mainly cock lifters, potato planters and potato diggers. As well as the forge they also had a corn mill. They also repaired machinery. My late mother often told this story about my grand father. A farmer brought in a piece of machinery for some minor repair. When the job was completed he enquired as to the cost. "Oh it is for nothing" said my grandfather. "That's too much" said the farmer, meaning it was too generous. My grandfather knew what he meant but being a bit of a wit responded immediately with "well, go where you can get it done cheaper".

Unfortunately, none of these Blacksmiths are in operation today. Their premises have been closed for years. Instead, we have "Mobile Farriers, "who will visit your farm. They come complete with anvil, fire etc., and can shoe your horse, or carry out surgical treatment to it's hooves.

Chapter 9

Fair Day

Main Street on Fair Day

The highlight of the social calendar in those days was the monthly fair day. This took place on the first Monday of every month. On that day the farmers came to town to buy and sell cattle, sheep, pigs, poultry, horses, corn etc. Each part of town had its own activity. Starting at the top end of Market Square on the right, as you look down from Masonic Hall you had the suck calves. These were mostly brought in by lorry by the dealers or jobbers having bought them in the South of Ireland. On the left around the Market House (now library) you had poultry, corn, potatoes etc.

Typical calf dealer

(On either side of Main Street proper the carts of young pigs were lined up. These were knows as suckers or bonhams (pronounced bonavs)).

The farmers preferred to sell their suckers direct to another farmer or feeder. However, sometimes they sold, without realizing it, to a tangler who then proceeded to retail the litter of say 10 to 12 pigs in ones and twos. They would often make a profit of Half a crown (12 ½p) 15c per pig. This annoyed the farmer, because the Tangler made more profit in 1 or 2 hours than the farmer had after feeding the sow and the pigs for months.

These continued on the two sides till you got to Murtaghs pub. Then we had 3 or 4 stalls of "Dublin Women" with second hand clothes. On the opposite side we had more stalls selling school copy books, pencils, holy pictures and screw drivers and various tools and knick knacks. Also on this side we always had Connell's, with their wooden feeding troughs, stools (milking), gates, wheel barrows, etc.

Around this area we also had in season a couple of men selling cabbage plants and a couple of stalls selling "burnt lime" for white washing. I should explain that in the fifties, the exterior of most houses were "painted" with a mixture of BURNT LIME AND WATER. To make "whitewash" you just added lime to water until the consistency was correct. If the colour was not white enough you added a "BALL OF BLUE". Whitewash was also used around the sheds in the farmyard where it was considered to have great disinfectant properties.

If there were a lot of pigs about they often continued on down on both sides after the NIB (then Northern) and Post Office to the corner of the "Old Green".

Next come the turn of the sheep. They were mostly left in horse carts but sometimes they would be taken out of the carts and tied with a hay rope to the wheel of the cart. On one occasion when I was about 10 or 12 years old I went to handle one of these sheep, she pulled on her hay rope very strongly, broke the said rope and took off down the street while I fled in the opposite direction. I didn't dare go down the fair again that day in case the farmer who owned her caught me. More about sheep later.

At lower Main Street we had the cattle. These were mainly Shorthorn, Hereford or Aberdeen Angus. Friesians, while in the country by the fifties, were very rare in Co. Cavan and Charollais, Simmental, Limousine and Salers had not yet been imported. These cattle had to be walked into town; if they came from further away they might have been on the road from 5 - 6 am. Nowadays they would be transported to the mart (stockyard) by tractor and trailer or by lorry. On arrival at the mart they would be put in pens until they were weighed and sold by auction. In the fifties they had to stand on the street and the farmer and or his son had to stay with them to ensure that they did not get mixed up with someone else's cattle.

If the cattle trade was bad or money was scarce you could stand about from 8am to 4 or 5 pm and not have anyone bid for your cattle. They were sold by "the lump" and it was up to buyer and seller to be able to estimate their weight and value. The "deal" could take anything up to an hour to complete.

For a 10 cwt (500KG) bullock (present value €650 to €800) the value in the fifties was probably £50. At the start of the deal the buyer would approach the seller, "How much do you want for the roan bullock?", the seller would ask £60 – the buyer "you are way out, I will give you £40". Maybe an hour later, after much hand slapping and bidding the buyer might offer the final bid, it could be anywhere from £48 to £52 depending on whether he or the seller was the better judge of the true value of the animal in the first place. To indicate that it was his final bid the buyer would take the right hand of the seller in his right hand and hold it palm up, then having spit on his own left palm he would slap the seller palm to palm. If the seller agreed to accept the price offered he would then spit on his palm and repeat the performance. The "deal was done" and the "spitting" was for luck. All that remained was the "Luck penny". This was a small amount of money that the seller gave to the buyer. It could vary from half a crown = 2 shillings and sixpence = 12½p = 15c to ten shillings = 60c. If the seller was smart he made sure to have very little change in his pocket.

If the cattle were bought by another farmer they would probably be brought straight home. However, if they were bought by a dealer they would be brought to one of the yards that were available for holding cattle while the dealer bought some more. The largest of these yards was at the back of what was then Kelly's Pub next door to the Model school. My friends and I would spend school playtime sitting on the wall being either "farmers or cattle dealers". By late afternoon there could be 50-60 cattle in Kelly's yard. The problem was sorting out one dealers cattle from another, as they were mostly the same colour. Nowadays all cattle have a yellow tag in their ear with a number which is unique to that animal. The same number appears on the "blue card" which goes with every animal. In the fifties one had to mark ones own. This was done either with a coloured raddle stick or with a scissors. The dealers had a special scissors with which they cut the shape of a letter on the animal's hair.

To continue on our tour of the fair, cattle continued on both sides of lower Main Street and along Church Street on the side of the houses, round the corner and up the Back Road. On the other side of Church Street – we had the horses, all along the Church wall. These were mainly heavy working horses, bearing in mind that in the fifties there were very few tractors and all the farm work, ploughing, tilling, mowing hay, transporting goods had to be done by horses. The largest horse fairs would be in   Sept – Oct when the three quarters and six quarters were sold (yearling and two year olds).

I can't remember any animals in Anne St. or Henry St. In Barrack St. we had all pigs other than suckers already mentioned; bacon or pork near Brady's corner and sows at the other end.

Then on the third Monday of the month we had the Middle Market. This was a much smaller affair. For the most part it was only lots of young pigs (suckers). Also on the second Tuesday of every month we had the pork Market. On these occasions an agent for one of the local bacon factories set up his weighing scales in Pat Brady's gateway in Barrack Street. On arrival he declared that he would be paying say £8 per Cwt (€10 for 50 kg). Each pig was then weighed and if the weight and price was acceptable the pig was turned into Brady's yard. If not acceptable you brought the pig home.

The stocking rate or the numbers of animals kept on each farm has increased vastly over the last fifty years. Kingscourt also had a monthly fair. Now the mart is held weekly and there would be twice as many cattle at each weekly mart as would have been in the combined monthly fairs in Bailieborough and Kingscourt. Similarly at Carnaross Sheep Mart weekly, there are more sheep than the combined monthly sheep fair of Bailieborough, Virginia, Mullagh and Kells. These extra numbers would be partly due to the farmers having improved the land so that it was capable of feeding extra stock. However, it is also due to improved breeding so that the cows produce a live calf every 365 days and the ewe produces and rears at least one and a half lambs per year.

I mentioned earlier about walking cattle into the Fair in Bailieborough. Any farmer who had to walk in the Virginia road had an additional problem. About a quarter of a mile out of the town, they had to pass "Terry the Butcher's slaughterhouse". Whether the cattle could smell the blood or just sense it, they would get quite agitated, and refuse to pass by. At the beginning of this chapter, I said that the Bailieborough Fair Day was on the first Monday of every month. Well, Mullagh Fair Day was always the last Thursday in the month. It always seemed to rain on the evening of Mullagh Fair. This was to wash away the smell – no, not the smell of the cattle, rather the smell of the herrings and other fish sold in preparation for "Fast Day" on the Friday!

Chapter 10

Preparing for Fair Day

For some people there was a lot to do before Fair Day. Anyone who served meals had potatoes and vegetable to prepare. I often helped in Grahams Hotel, when up to 100 kilos of potatoes had to be peeled on Saturday and kept in large galvanised baths of water. The same quantity of carrots, turnips etc had to be prepared.

The shoe shops were also busy on the Friday, Saturday before the Fair cutting leather. In the fifties all shoes had leather soles and for the most part these were repaired at home. The leather came in a "side" approx 3' by 6' and had to be cut up into soles or heels etc. My father had a special steel ruler with a piece of lead inserted into one side for marking the leather. When this ruler was pulled across the side of leather the lead left a mark showing where to cut. There was a special knife for cutting the leather. The side of leather was thicker in some parts than other, depending on what part of the cows hide it came from. Leather was sold by weight, so when the soles, heels etc. were cut they were weighed and priced. The customer would then select a light or heavy piece depending on whether it was for a pair of heavy working boots or light Sunday shoes. We also had to weigh out "sprigs" into 1oz packets for nailing the leather soles onto shoes etc. In the same department we had steel tips, toe pieces and studs to complete the job. Speaking of footwear, in the late forties there were country folk who still wore clogs. Now the Irish clogs were different from those worn in Holland. The Dutch ones were made from all wood, whereas the Irish ones had a wooden sole but leather upper. My father used to sell the replacement wooden soles for these.

Working boots have also changed. The best working boot was made from Kip Leather. A slightly lighter version was the Split Kip. I can still remember the wonderful smell of kip leather. These boots were not tied with laces but rather with whangs. With what, I hear you ask? Whangs were strips of Kip leather about the same thickness as present day laces. Again, a lovely smell when new. Another form of footwear that we never hear of nowadays are "galoshes," these were made from black shiny rubber. If you were going to Mass or Church on a wet murky morning, you pulled these on over your shoes. When you got to your destination you took them off and your good "Sunday Shoes" were still clean and dry.

Another task before the fair for anyone with a large yard was to tidy the yard in preparation for the stabling of the horses from the country on Fair day. Graham's yard was one such. On a busy fair day there could be up to 30 animals, horses, ponies and donkeys stalled there. Often there were more horses etc than stables for them so the excess had to be tied to posts, gates or anything available. There was a charge of 6 old pence (approx 3 cent) per animal, which was collected by the yardman (in this case Paul McDonald, father of John Paul, Painter and Decorator) rather than the yard owner. After all it was the yardman who had to sweep up after them. If you consider that a full weeks wages at that time was probably £4, this 15/= (shillings) extra was a nice bonus.

Some shopkeepers also fitted wooden barricades around their shop fronts, in case a bullock tried to jump into the window.

Chapter 11

Transport Regulations

I said at the beginning that I would come back to Willie Roundtrees Lorry. In the early fifties all the public road transport in this area was carried by G.N.R. (Great Northern Railway). In Cork it would have been G.S.R. Early in the sixties C.I.E. (Coras Iompar Eireann) came on the scene. At the same time a few private Lorries were beginning to appear. C.I.E. was worried that these private lorries would slow down their growth so they got the Government to require lorries to have haulage licences. Now, if I owned a lorry I could transport my own cattle from here to Cork, if I so wished. However, if I was required to transport livestock for another farmer for reward I must have a licence and then I could only operate within County Cavan.

On the morning of my story, I was talking in the sheep fair to a sheep buyer from Crillys sales yard in Dundalk. Later in the afternoon I was standing in the door of our shop when I saw the same buyer coming up the footpath. It was obvious that he had a problem. I spoke to him, enquiring what the problem was. He was looking to get access to a telephone (this was before mobile phones). It transpired that he had just received a telegram to say that his own lorry, which was coming to bring home the sheep he had bought, had broken down and could not make it to Bailieborough. He would have to hire a replacement. Now, as I said, Willie Roundtree's lorry was sitting along the courthouse wall but as he was only licensed for Co. Cavan he could not go. There was another lorry belonging to Willie Copperthwaite parked at the other end of town, same story.

I brought the "sheep man" into our phone. He rang C.I.E. in Carrickmacross, no lorry available, Dundalk same story, Kells similar. He eventually had to hire a lorry from C.I.E. Navan, to come to Bailieborough, (40kms), load up the sheep, drive to Dundalk (another 40kms) and then return to Navan (80kms.) This, with two lorries capable of doing the run, sitting in Bailieborough, but could not go due to red tape.

Nowadays CIE no longer transport livestock, either by road or rail. If Willie Roundtree or Willie Copperthwaite was around today, either could transport these sheep anywhere they needed to go.

Most farmers when they need to transport their livestock to the mart, or to the meat/bacon factory, they either use their tractor and trailer or 4-wheel drive and trailer. However, if a farmer does not have his own transport there are "Hauliers" around to do the job.

Chapter 12

Veterinary Service

A contributing factor to the increase in stock numbers would be the improvement in veterinary services both from the training of veterinary surgeons and also in the range of "drugs" available to them.

In the late forties and early fifties we had one Veterinary Surgeon in town the late "Vit" Donoghue, who lived in Church Street. Then the late Willie McKinley came home from Northern Ireland to set up practice here . When Donoghue died the late Jim Sheridan, father of Conor, came to town. Both Mc Kinley and Sheridan had many assistants down through the years. However, three in particular from the Mc Kinley Practice stick in my mind. The first was a man of about 50. He was a very nervous driver and it used to take him twenty minutes to drive the length of Main Street (200 yards) and I can assure you that the traffic then was a lot less than now. The second was a young man just after qualifying. He stayed in Grahams Hotel (The Bailie). I remember, one cold February night Roy Graham had a sow about to farrow (pig). As Roy had the "flu" I volunteered to "sit" with her. Before the Hotel staff went off duty I enquired as to which room the vet was in, just in case I might have need of his services during the night. At about 10 o'clock the vet wandered up the yard to see how things were going. The sow had not started at this stage. We talked for a while and it became obvious that the vet did not know whether the pigs were going to come out of her mouth, her ear or under her tail. He came back about midnight. The sow was in full swing and he started counting – one, two, three...six, seven. "Goodness isn't she having a lot". If he had come back again at one o'clock and saw fourteen piglets running round, he would have lost his life. Fortunately, I did not have need of Veterinary assistance that night, because I don't honestly think he would have been much use to me. I don't know where the latter qualified but at that time ever second or third call to Mc Kinleys office would have been concerning pigs. Needless to say he did not last long in County Cavan.

Willie Mc Kinley also employed the first lady vet around here. At first the farmers were a bit wary of her, but after a while she became accepted and they would ring up and say to Willie "don't bother coming yourself, just send the young lassie".

This came about in the following way. On the first occasion when she was sent out to neuter a litter of pigs, the neighbours all gathered round to witness the performance. She was half way through when one of the spectators said "be gob miss, it's a terror to see a young lassie like you doing a job like that". Quick as a flash she grabbed hold of him with one hand while still holding the knife in the other. She said "if you are not careful I will perform the same operation on you". Needless to say there was a deep silence in the pigsty for a while!

In the fifties not many people in the country drank coffee, and so not many people used Demerara sugar (brown). On one occasion a farmer visited Mc Kinley's surgery with a pig that was constipated (bound). Mc Kinley examined the pig and ascertained that there was nothing seriously the matter. He told the farmer to go up town and get a pound of Demerara sugar. The farmer had never heard of this, so when he arrived in Roy Grahams shop he asked for a pound of diarrhoea sugar.

In the late sixties, early seventies another vet, Paddy Prenderville came to the locality. He set up practice in Kingscourt, 12 – 13 km from Bailieborough. Paddy has now left Private Practice and works for the Department of Agriculture. When he heard that I was writing this book he suggested that I include the following two stories.

When he came to Kingscourt first Paddy frequently acted as "Locum" for both Mc Kinley and Sheridan. This would often be at night or at the weekends. On one occasion, late at night he was sent on behalf of Mc Kinley to visit a farmer who had a very sick cow. After a careful and thorough examination of the cow, Paddy decided that there was nothing he, or anyone else for that matter, could do for the cow. He advised the farmer that if the cow was still alive next morning he should arrange to get her into the local abattoir.

Next morning the farmer decided, seeing that the cow had survived the night, he would seek a "second opinion". He rang Sheridan's Office and requested that a vet should come out to examine the cow. However, he did not disclose that a vet from Mc Kinley's Office had been out the previous night.

By sheer coincidence, Prenderville happened to be covering for Sheridan that morning and duly arrived out. Needless to say the farmer was not impressed to see Paddy arrive into his yard again!

The second story also happened at night, only this time it was a very stormy one in mid-winter. The call out was to a sow that had been in labour for a considerable length of time but was unable to farrow (have her litter). Paddy was in the process of giving the sow an internal examination when there was a loud clap of thunder and flash of lightening. Next thing the lights went out and the two men and the sow were plunged into darkness. In an effort to lower the tension the farmer said "Where was Moses when the lights went out"? Paddy replied "Well one thing is certain; he was not lying on the floor with his arm up to the elbow in a sow's ....!!!!

To conclude this chapter, Willie McKinley left the area in the late sixties, or early seventies. He is now deceased. Jim Sheridan is also deceased, and his practice has been taken over by his son, Conor. We now have a new vet in the district, Philip Lynch.

Chapter 13

Dairy Cows

In the forties and fifties most of the Dairy cows around here were Shorthorn. Originally we had two distinct types of Shorthorn, Dairy Shorthorn and Beef Shorthorn. As the names would suggest these were bred for two different purposes. Then the "powers that be" decided to cross the Dairy Shorthorn with the Beef Shorthorn with the intention of coming up with a dual-purpose animal. The idea was that this new breed would still give a reasonable quantity of milk and give a good calf as well. Unfortunately it did not work, in fact it sounded the death knell for both Shorthorn breeds. By the 80's there were very few Shorthorn herds left in the country. Fast forward to 2006 and you would have difficulty finding one decent Shorthorn herd.

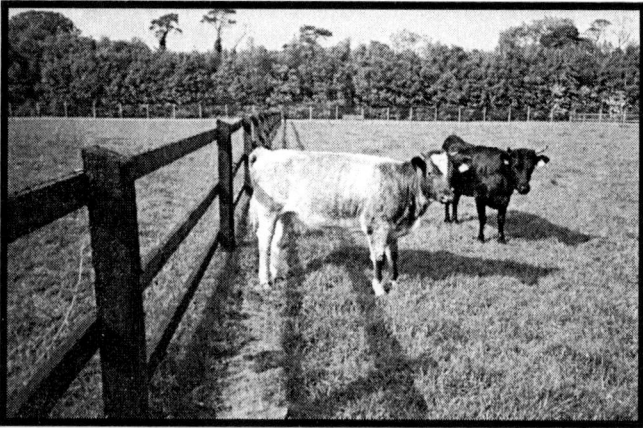

Typical strawberry roan shorthorn heifer, with native Kerry heifer behind

Rewind to 1940. As I said, mainly Shorthorn, a few Jerseys and a few Friesian. The British Friesian was first imported to Ireland in 1910 /20 mostly into counties Cork and Limerick. Two of the first Friesian herds in Co. Cavan were established in the late 40's in Arva by Hugh Hourican and Con Bouchier. One of the first herds in Bailieborough, as I said earlier was set up by Roy Graham. This would have been in the early fifties.

The main attraction of the Friesian over the other breeds available at that time was their ability to produce large quantities of milk. At that time a top Dairy Shorthorn cow would probably give about 400/450 gals of milk per year. The British Friesian on the other hand could give 700/750 gals. Unfortunately the Friesian at that time had a very bad reputation for butterfat. The butterfat content of the milk was so low that it was sometime referred to as "Blue milk". However, with dedicated breeding in a few years both yields and butterfat were improved quite remarkably. It is ironic that by the nineties farmers were being penalised for having too much butterfat. This is still the situation today.

An indication of how few Friesian were around in the mid fifties was the fact that Virginia show did not have classes for Friesian at that time. When Roy Graham entered two heifers in 1955/56 they had to go into the class for "Dairy Heifer other than Shorthorn". In this class there was a mixture of Friesian, Ayrshire and Jersey. My brother and I exhibited these heifers for Roy.

Similarly when I went to Gurteen Agricultural College (on Offaly/Tipperary border) in 1957, the college dairy herd (approx 40 cows) were all Shorthorns.

However, when I got my first proper job in 1959 on Saint Edmundsbury Hospital farm, at Lucan, Co. Dublin about 1/3 of the herd were Friesian, the remainder being Ayrshires. They had 75 cows in total.

During the 1950's farming in other parts of the country was beginning to change and we started to see the start of mainly single enterprise farms. Cork, Limerick, and Tipperary (the Golden Vale) concentrated on milk production, Meath and Kildare tillage and so on.

However, Cavan continued with mixed farming right up to the beginning of the seventies. By mixed farming I mean they milked cows (15-20), kept some dry cattle, maybe 10-12 ewes, two or three sows and a selection of poultry as described in another chapter. They also grew perhaps 2 acres of potatoes, 1 acre for their own use and the other as a cash crop. A small area of oats was also grown to provide winter feed for their livestock.

It is easy to see why they did not have too much livestock around the yard. There was no electricity and therefore no water on tap. A milk cow drinks 15-20 gallons of water per day, a sow and litter of pigs 4 to 5 gallons and all this had to be carried from the well, which might be twenty yards away. Also if you had to go out to the yard after dark you had to bring an oil lamp, known as a storm lantern.

In the early sixties things began to change. This was due in no small way to Rural Electrification. One of the first things that most farmers purchased was an electric water pump. As well as having running water and flush toilets in the house they could have running water all around the farmyard. The next major item was a milking machine. New cow byres were built and cow numbers increased. The first type of "Milking machine" was what was known as a "bucket unit". With this you had a vacuum line going down one side of the cow shed or byre and back up the other. On this line there was a "tap" between every two cows. You came along with your "bucket unit" and plugged it into the line at the tap. When one cow was milked you put the machine on the other cow. When both were milked you unplugged and moved to the next pair of cows. It was reckoned that one needed approximately one-bucket unit for every six to eight cows to be milked. The big problem with the bucket plant was that the milk had still to be carried back to the dairy. This problem was overcome with the development of the pipeline system. More about this later.

About 1964/65 I met and spoke with a man who had six cows. This case would scarcely justify a milking machine at all, but he bought two units. When I asked him why he bothered with the machine at all he said "well, I am not getting any younger and I am not as fast at the milking as I used to be". He then proceeded to put the machine on two cows and sat down to hand milk the third cow. And then the same with the other three cows, making sure that the cows that were hand milked in the morning were machine milked that evening and vice versa. When asked why he did not milk all with the machine he replied "If ever there was a power failure the cows would have forgotten what it was like to be hand milked and also my hands would be out of practice as well".

Meanwhile the numbers of Friesian were on the increase. Here locally they had a staunch supporter in one Barney McCabe B.Ag., an agricultural advisor with Cavan Committee of Agriculture stationed in Bailieborough. One talks about young lovers seeing each other through "rose tinted spectacles" well Barney saw things through black and white tinted glasses. Often he would be called out to advise on a batch of say Hereford bullocks but if he saw Friesian cows in the next field he lost interest in the bullocks. As far as he was concerned, Friesian was the only breed of cattle worth feeding. By the early 70's one was beginning to hear of Friesian cows that had produced 1000 gallons. By this time Barney McCabe had been replaced by the late Tim Ryan B.Ag. At a farmers meeting in the Bailie Hotel one night Tim was asked if he thought that the 1000 gal cow would become commonplace. Tim thought for a while, then said "forget the 1000 gal cow and think about the 1000 gal man". In other words the farmer had to be mentally geared to feeding and managing his cows with 1000 gals in mind, otherwise it would never happen. Also at this time there was a debate going on as to whether one should have heifers calving for the first time at three year old, which was normal at that time, or could she be calved down at 2 years old. At the same meeting as above, Tim was asked "what age should you mate a heifer for the first time." This time Tim answered "when she is big enough she is old enough". He went on to explain that a heifer needed to be 6 hundred weight (300 kgs) at time of mating, in order to calve down at 10 cwt (500 kgs). If she had been adequately fed as a calf and over her first winter she should be 300kgs at fifteen months of age and would therefore calve down at 2 years old. If however she did not reach 300 kg till 2 years old, so be it, and she would then be nearly 3 year old when she calved down. His words are as true today as they were 30 -35 years ago.

One major change took place to the Friesian breed in the late 70's early 80's. This was the introduction of Holstein blood from America and Canada. As a result of this, yields began to rocket but butterfat and more importantly protein levels were also vastly improved. At the present time (2006) a good average herd yield would be 1500-1600 gallons.

However, 2000-2500 is becoming more common and occasionally one hears of 3000 gallons in a 305-day lactation. The interesting thing to note is that the farmers who are achieving those higher figures are sons of the men Tim was talking about all those years ago.

However, 2000-2500 is becoming more common and occasionally one hears of 3000 gallons in a 305-day lactation. The interesting thing to note is that the farmers who are achieving those higher figures are sons of the men Tim was talking about all those years ago.

In case my readers would get the impression that the Irish national dairy herd consists of nothing but Friesian let me set the record straight. There are still pockets of Shorthorn, Ayrshire, Jersey and even our native Kerry. During the past twenty years we also had importations of several new breeds. These included MRI, Rotrundt Red, Swiss Brown and Montbeliarde. All have their merits and staunch supporters.

To go back to our milking machines. The pipeline system was where only the actual milking unit was moved from cow to cow no need to carry a bucket around. The milk was drawn straight to the dairy. The next move was to the "milking parlour" where the cows were brought to the machine rather than the machine to the cows. This reduced the milking time and allowed cow numbers to be increased.

Milking parlours got more and more sophisticated over the years. By the mid nineties we had automatic cluster removal and computerised feeding. Then in the late nineties we got the Robot milker. The cows could come in or out whenever they wanted and the robot would feed them, milk them and put them out again. However, due to cost I can't see this system being in general use for a long time to come. On the other hand, the five-day cow may be getting nearer.

Some farmers are now only milking 13 times a week instead of 14. In other words, they only milk once on Sunday. Others practice O.A.D. (once a day) for some weeks in springtime. Who can say what changes will take place in the next 10/15 years?

Chapter 14

Beef Cattle

As I said earlier the main breeds of beef cattle in the fifties were Beef Shorthorn, Hereford and Aberdeen Angus. These were mostly sold as Stores at 11/2 – 2 years old. Very few cattle were finished locally and instead were bought by Graziers in Co. Meath where the grass was better. The biggest cattle fairs of the year were in the spring time and again in the autumn. The reason that these Spring Fairs were big was because the farmer wanted to reduce stock numbers. He might have last year's cattle and the years before. He therefore had to reduce stock in order to save grass for this year's calves. Similarly in autumn if the summer had been bad and the quality and quantity of the hay saved left a lot to be desired, stock had to go. This was also true if the signs pointed to a bad winter on the way. I should point out that in the fifties most farmers only had houses for their cows and young calves. The other stock had to remain out over the winter.

In the 1950's very few cattle were produced initially for beef. They were in fact a bi-product of the Dairy Herd. Even those who now kept Friesians used a Hereford or Angus bull. The calves were bucket fed and eventually made their way into the food chain. In those days it normally was three years before these cattle were ready for slaughter.

Once again the arrival of electricity to the countryside brought about big changes. I spoke in the previous chapter about the Dairy Herds getting larger. This in turn meant more cattle for beef. The farmers then started to build sheds to accommodate these cattle. Having these sheds meant that more cattle could be retained for longer on the farm of birth, which in turn meant that more cattle were finished locally.

During the sixties and into the seventies three things happened that changed the production of beef in this country for all time. The first of these in about 1962 was the initial importation of Charolais cattle from France. This was followed two years later by the arrival of the Simmental. The Limousin followed a few years later. Other breeds have followed since – Blonde Aquataine, Romangola, Belgian Blue, Salers and Aubranc.

Most breeds of cattle have their own characteristics. Let us start with the Hereford. Often known as the "White head", a pedigree Hereford has a dark brown or red body, all white head with the exception of the ears which are brown, a white stripe running down its back and a white tip to its tail, and quite often but not always four white socks. No matter what breed of cow is mated to a Hereford, the calf will always have a white head, and the body is normally brown. However if the cow is Friesian, Aberdeen Angus or a Kerry, which are black, the resulting calf will have a black body, black being the dominant gene.

The Aberdeen Angus is all black, and is naturally polled, in other words it never grows horns. Regardless of the breed of the female all calves born from an Aberdeen Angus sire will be all black and polled. The exception to the colour rule was when a Shorthorn cow was mated to an Angus bull; the offspring was often what was known as a Blue Grey. These were quite common in the West of Ireland and were much sought after by farmers from Scotland.

Because of their size (small), Aberdeen Angus bulls are often used on heifers in the dairy herd. This allows the heifers to be calved down at 2 years old without any strain or pressure being put on them at calving time.

Charolais: Here the pedigree animals are pure white. When crossed with most other breeds, the offspring are normally Blonde or Tan. However when mated to a Friesian, the calf is normally mousy grey. This is because the dominant black gene is trying to break out.

Simmental: Colour marking of the Simmental is very similar to the Hereford with one exception; in the pedigree animals, while the head is white like the Hereford there is normally a ring around the eye of brown pigmentation. However this eye pigmentation was not always passed on in the crossbred calf. Personally I feel this is why as a breed the Simmental did not get as much prominence as the Charolais in its early days in Ireland. I should explain that when the dairy farmer brought his calves to market at 2/3 weeks old he always got a premium for "Continental calves", for example in the sixties a Hereford male calf was probably worth £25 but a Charollais or Simmental (if bearing eye pigmentation) could be worth £40 or £50. If the eye markings were not there a Simmental at that age looked just like a Hereford and the farmer was paid accordingly. However if we go forward one year when that calf begins to grow and fill out the continental blood begins to show. The animal would be much bigger than a Hereford of similar age and at maturity could be anything up to 50 – 100 kilos heavier.

Incidentally, when the Simmental were first imported they were promoted as a dual purpose breed, because of their ability to produce milk but also to produce a calf that would grow into an excellent beef animal. I know people who have milked ½ bred Simmentals where the sire was Simmental and the dam Friesian and they have made excellent milkers.

The Limousin are brown in colour and the cross bred calves are normally bright brown or black depending on the breed of the dam. In 2006 they are probably the most popular Terminal bull for a suckler herd.

As the name would suggest Blond D'Aquitaine are blond in colour and the crossbred offspring are usually the same. They were known for their double muscling.

The Belgian Blue can be white, blue-grey or black and white, and the offspring somewhat similar. Because of the size of the calf at birth they have often to be delivered by caesarean section. In fact in their native Belgium the pedigree animals are automatically sectioned when they come to their time.

I would like to go back to the Charollais/Simmental. In the late sixties the Breed Societies for both these were anxious to increase the number of Pedigree stock in the country. At the time there were restrictions in place with regard to importations so they decided on a scheme of "upgrading". You started off with a Pedigree Charollais bull mated to lets say a Friesian cow. The female calf from this mating would be graded as a half-bred provided it had the appearance of a Charollais and would be issued with a certificate stating it was registered as a half bred Charollais. In due course you mated that female to a pedigree Charollais. Now the female calf, all things being equal was registered as a three quarter bred; use a Charollais again, and the female calf of this mating would be seven eighth bred. Next generation would be fifteen sixteenth. One more generation and you had full pedigree. The same scheme applied to the Simmental.

About 1969 I decided to have "a go" at the grading up. I went to an auction in Kilnaleck and bought an in-calf registered ½ bred Charollais cow. In due course she calved a bull calf – no good, the scheme only progressed on female calves. Next year –same story – bull calf and believe it or not the third year, bull again. I gave up after that.

A cousin of mine was also trying the grade-up route, he had both Charollais and Simmentals. He advanced further than I did, but never got as far as full pedigree. However, a few years later he bought his first natural Pedigree Charollais and now keeps nothing but pedigree.

I also bought pedigree, but I went for Simmental. However, I only stayed at pedigree for a few years, as I had to sell them to raise capital when I was setting up my mushroom business (more about mushrooms later).

I said earlier that there were three things that changed beef farming. The second and most important was in 1973 when we joined the then EEC (European Economic Community). This opened up vast new markets to Ireland. There have been a lot of changes since then, some good, some bad, but personally I am convinced that if we have not joined Europe

49

in 1973 we would be a much poorer country today. I don't think the "Celtic Tiger" would ever have roared.

It has not been all sunshine; most farmers of my age remember the disaster of 1974. I cannot remember the details, but suddenly the price of cattle dropped through the floor. This was particularly so in case of young calves. Good calves were been sold for £1.00 - £1.50 but the poorer quality calves could not be given away. I heard of many cases where a farmer brought one calf to the mart, but when he returned to his trailer he found that four of five calves had been left in it. Fortunately this hiccup did not last for very long.

I should have mentioned the FARMERS MARCH that took place in 1966. On this occasion farmers from every parish in Ireland walked or marched to Dublin. On arrival in the capital they held a massive protest rally. At this rally the farmer's leaders spelt out their grievances. After the rally a number of farmers were selected to camp out on the footpath outside Leinster House (Government Buildings). They set up tents, gas stoves etc. For weeks they camped out day and night. Some of the protesters were arrested and spent some weeks in Portlaoise Prison. As one farmer was arrested, there was at least one more ready to take his place. At this point the GOVERNMENT realised the farmers were not going to "curl up and roll over". They agreed to meet the farm leaders provided they called off their "sit in". Thankfully, relations between the POLITICIANS and the farmers have never sunk to as low a level since.

Back to beef farming. The third important change that took place was the advent of PART-TIME FARMERS. A lot of young farmers in the late sixties realized that their farms were too small to give a living wage similar to the "Industrial Worker". They were not prepared to slave away for very little return as their fathers and grandfathers had done. The solution was to get "OFF FARM EMPLOYMENT". Once settled into this system they were able to run the household on their "INDUSTRIAL"wages, and so could reinvest the farm profits into improving farm buildings, buying better quality livestock and availing of modern technologies and labour-saving devices. Some of the most progressive farmers in the country at the present time are to be found in the ranks of these "PART TIME FARMERS".

Chapter 15

Poultry

In the 1950's every farmhouse yard had a few hens, ducks, geese etc. roaming around the yard. In 2006 very few houses have poultry, but those that do, have thousands rather than a few.

Sometime in the late forties, early fifties, the then Minister for Agriculture, I think it was Mr Dillon, said that Ireland would flood England with eggs. To this end he gave grants to farmers to build new hen houses. These were all the same design and could be seen dotted all over the countryside. Prior to this, in most cases hens slept in the rafters of the stable or cow shed or in the corner of the cart shed. This meant that the eggs were laid anywhere and everywhere. When the new hen house was built it meant the hens had a regular place to roost and lay. They could also be shut in at night, which meant that the fox did not get as many of them. A lot of those hen houses are still about but it is a long time since some of them last heard a hen lay an egg.

The problem with egg production in the fifties was that because of the short days and long nights in winter the hens only laid for about eight or nine months of the year. In order to have plenty of eggs in November/December for the Christmas baking the housewife had to preserve eggs when they were plentiful in summer. To preserve eggs you had to butter the eggs with a mixture of lard and borax powder or alternatively you placed them in a jelly like substance called water glass. When the eggs were treated in this way they were placed in a cool dark cupboard to be taken out and used as and when required over the winter months. These would not have been fresh enough for boiling or frying but were excellent for baking. The only problem was, that if an egg was not as fresh as it might have been at the time of preserving, when it was cracked months later, the stink could remain about the house for days.

Back to our hens. In the fifties these were mostly Rhode Island Reds, Light Sussex, White Wyandotte, White Leghorn and Minorcan Black . I can remember a joke that was doing the rounds when I was at school. "Why is a black hen better than a white hen?" answer "A black hen can lay a white egg but a white hen cannot lay a black egg"!!

Nowadays these breeds have for the most part disappeared except for a small number of specialist bird fanciers who breed them for showing. The modern hens are mostly Hybrids or mixtures of a number of the old breeds and are identified by number, each hatchery having their own "brand".

One of the biggest problems with hens in the past was "broodiness". This was when hens became broody and went and layed their eggs in some obscure and hidden place, where the farmer could not find them. The hen, having laid 6 or 8 eggs in this secret nest would sit on them day and night to keep them warm, only rising once per day to return to the farmyard for food and water. A few weeks later she would reappear with 6 or 8 chicks in tow, provided that the fox, the rat or the stoat had not got to them first.

Once the chicks appeared the farmer had no choice but allow her rear them. This meant that while she was on mothering duties she was not laying eggs. When the chicks were 5 or 6 weeks old they were taken from the hen and put with the other "young stock". The females or pullets were kept to become layers and the males or cockerels were fattened for the roasting dish.

Back to our layers, in the good old days the eggs were mostly sold to travelling shops, like Matt Traynor in exchange or barter, for tea sugar etc. Others had their regular customers in town that they supplied every week.

In my earliest recollection, replacement birds were hatched at home – see earlier paragraph, but by the late fifties, early sixties people were beginning to order them from the local hatchery, like Elmbank in Cavan. They could be ordered "as hatched", in which case you got a mixture of male and female, or you could order "pullets only". These were despatched from the hatchery in cardboard boxes with ventilation holes provided, I hasten to add. When travelling by bus in those days it was a regular occurrence to hear the "cheep cheep" of "day old" chicks on the back seat. This was the common method of transportation.

A major change took place in all areas of farming with arrival of Rural Electrification in late 50's early 60's. This was also true in the poultry sector. With electric light in the "hen house" it was possible to increase the hours of light and therefore keep the hens laying for most of the year. The first major change was that the hens were taken in from the yard and shut in "deep litter" houses. As the name suggested, the house was only cleared out once per year, however, fresh litter was added each week. This fresh litter gave the hens something to scratch through to relieve boredom. In those deep litter houses the electric lights could be put on early in morning and again in mid afternoon. This tricked the hens into thinking that the day was much longer than it was and so they spent longer laying.

The next development was battery cages. Apart from anything else this meant that more hens could be fitted into the same space. In a house that would hold 100 hens under the deep litter system you could put 1500 – 2000 in cages. This was really the start of factory farming. Prior to the advent of "Deep Litter" and "Battery" a flock of 100 hens was considered very large. Suddenly with battery cages we are talking telephone numbers 10,000, 15,000 even 20,000.

In the sixties, International Hatcheries like Cobb and Whittakers set up branches in Ireland. These hatcheries were constantly striving to "breed" birds that would lay more and more eggs. Some produced "white egg layers" others had "brown egg layers". In the mid 60's a very large egg producer, whose name escapes me at present, started advertising on T.V. that his brown eggs were better than anyone else's white eggs. He claimed that the eggs had more proteins, vitamins etc. Now I was involved with the poultry industry of the time and to the best of my knowledge the only major difference, apart from colour, was that in general the hens that layed white eggs consumed on average 3½ oz of meal per day, whereas those that layed brown eggs ate approx 4 oz per head per day. However, his advertising campaign obviously got through to the general public, because to this day some housewives will buy nothing but brown eggs.

The advent of these large factory farms meant the demise of the small farmyard units. Some farmers continued to keep a few hens, just to produce for their own household, but soon that died out too as eggs could be bought cheaper than they could be produced. However, let us fast forward to 1990, people are on a health kick and "free range" eggs are back in fashion. At first everyone was jumping on the bandwagon and there was a lot of confusion as to what was free range and what was not. By now everything was controlled by "Brussels". The Beaurocrats brought out rules and guidelines to which one had to conform. They specified the size of opening required to allow the hens to get out but if only 2 or 3 hens come out everyday you were still entitled to sell your eggs as "free range". Come forward another 10 years and the small flock for ones own use is coming back into vogue. This is especially so in the case of families moving from Dublin to live in the country.

Before leaving "Egg Production" I should mention Bantam Hens. To the uninitiated they are about half the size of normal hens, are very colourful and lay eggs that are much like hen eggs only half the size.

Enough about egg production. Let us move on to "Table Poultry". As I said earlier this was mainly produced from the birds that were not required for Egg Production. In the fifties a few people tried to operate special table bird units. This was called "Capon" production. Capons were neutered cockerels. They were shut in all the time and were force-fed on whole grain. The system did not catch on to any extent, mainly because the birds produced were too large and too expensive for the average housewife.

There was also another type of Table Bird produced called "Boilers". Because of the age at which they were killed, (four to five months) they were too tough for roasting and so had to be cooked by Boiling.

Once again the advent of Rural Electrification brought vast changes to Table Bird Production. The arrival of Carton Brothers Shercock, Cootehill Poultry Products and Monaghan Poultry Products in the Mid 60's really got the Broiler Industry off the ground. Wooden Pre-Fab houses sprang up all over the place and once again we were talking telephone numbers, 10,000 per house and maybe 3 or 4 houses per farm. These companies controlled everything – they supplied the day old chicks, either supplied the feed or told the farmer what feed he could use, any drugs or medication could only be used in consultation with the company, and finally, they decided when the birds were ready for slaughter, usually at 8 to 10 weeks. The birds were delivered to the supermarkets frozen. These companies still control the industry today.

Ducks
Quite a number of farms in the fifties kept a few ducks, especially if there was a river or stream flowing near the yard. These were kept both for egg laying and for table birds. The duck egg is slightly larger than a hen's egg. Personally I prefer them boiled to fried, and in my youth when asked how I liked my duck eggs cooked, my answer would be "beside another". Yes! Two boiled duck eggs would do me very nicely. Thank You! In the fifties there was little or no commercial duck production, either egg or table. Then in the early sixties all this changed in Emyvale, County Monaghan. This was when Silver Hill Farm was set up. When driving through Emyvale there were several fields that were as white as snow with all the white ducks. Nowadays most of the ducks are reared indoor. Silver Hill is the largest duck producer in Ireland and export their products all over the world.

Turkeys
Here again a lot of farms in the fifties kept a few turkeys. For the most part it was just to produce a few turkeys for their own and extended family at Christmas. A turkey egg is

about the same size as a duck egg and the shell is light brown, with speckles of darker brown thro it. It is so long since I last ate a turkey egg that I cannot remember what it tasted like. Any turkeys that were sold helped to provide the farmers wife with "pin money". I can still hear the noise of the turkey market on Market Square on the two Mondays prior to Xmas. The birds were brought to town in Horse carts or later in car trailers. They were sold alive. The purchaser brought it home alive and then made the necessary arrangements. While it was mostly turkeys in these markets, one also got geese, ducks and hens.

In the late 50's early 60's there was only one breed of turkey in Ireland, namely the American Bronze. This bird was descended from and still resembles the wild turkey to be found in certain parts of America. In the sixties we got a new breed – Broad Breasted White. As the name would suggest, they had more flesh on them than the bronze.

This was the start of commercial turkey production. At first turkeys were available at Easter as well as Christmas. Then larger units were set up to produce turkey all year round. In the eighties Turkey and Ham was the menu of choice for most weddings. Those producers who were in business all year round were for the most part on contract to a packing plant, like Cartons of Shercock. For farmers and or their wives producing for pin money at Xmas, things began to change. The "Live" market continued till about mid 80's. Then, for this area most turkey sales were transferred to the Cattle mart in Ashbourne. The birds were sold N.Y.D. (New York Dressed). This meant that they were dead and plucked but the head, feet and intestines were all intact. If you wanted a turkey "oven ready" you had to go to the Butcher shop. When Ashbourne mart closed around 1990 the turkey market transferred to Ballyjamesduff Mart. At that time I worked with Farm Relief Services and would be contracted by F.R.S. to the mart for "Turkey day".

In order to get an early number and therefore the best prices, farmers would start queuing up on Sunday evening (Monday market) and would have to sit in their car or van all night. It was a long cold night, especially if the turkey were in the car/van rather than in a trailer. It meant they could not start the engine, or put on the heater as this would affect the birds. When the Mart opened on Monday morning, the turkeys were weighed individually, and then sold by auction, just like cattle, sheep etc.

All this came to an end about the year 2000, when the Department of Health and Safety declared that this was not a suitable way to handle food products. While in 2006 the sight of farmyard turkeys is less common than it was, there are still some people who rear 20 – 25 turkeys for Xmas. They now have to be sold direct to the consumer.

Geese

In the fifties some farmers kept a few geese. A goose egg is the size and shape of a small rugby ball, rather like an ostrich egg. I must confess I never had the opportunity to eat a goose egg, but I know those who did. Some people would not thank you for turkey on Xmas Day. It had to be a goose or nothing. In flavour and texture goose flesh is to turkey what duck is to chicken, the flesh is coarse and more oily or greasy. Another difference between turkey and goose is that while turkeys, or chicken for that matter, are normally stuffed with a bread stuffing goose always had a potato stuffing. In fact to me eating this potato stuffing was the high light of the meal. I have often asked but no one can supply fitting answers as to why potato stuffing cannot be used in turkey or chicken. As I said goose has a lot more grease. This grease or "grace" as it was known was much sought after. It contained great healing powers; it was rubbed into the joints to relieve aches and pains. It was also beneficial in the treatment of chest ailments such as asthma and bronchitis. The feathers of the goose were also sought after for stuffing pillows or Eiderdowns. What's' an eiderdown, I hear you ask. Well, it is what we used to call a duvet, before the word duvet was invented. Some people also kept geese, for security purposes, they kicked up such a racket with constant gaggling if a stranger appeared that the entire household was alerted. Unlike chicken, turkey and duck, geese have never been farmed commercially (to the best of my knowledge). For the same reason I have never seen goose on a Hotel or Restaurant menu. A goose strutting around the farmyard is a rare sight today.

Guinea Fowl

Even in the 50's these were rare but some people kept them. Their egg is about the size of a Bantam egg. No, I never ate one. Again they were often kept as a security alarm or early warning system. When disturbed they could be heard from a long distance away, their cry of alarm sounded like "Go back", "Go back". I had not seen or heard Guinea Fowl for many years. However, on a recent trip to New Zealand, I saw and heard a number of them in Staglands Park near Upper Hutt.

Peacocks

These were often kept as a status symbol. I have to confess that I never saw a peacock's egg – for the very good reason that the peacock does not lay eggs – no, it is the peahen that lays the eggs. For the benefit of those who have never seen one, the adult peacock is about the size of a turkey. The body feathers are quite colourful but it is really the tail feathers that are the crowning glory. They are about 3 to 4 feet long, have circles of colour, usually green starting with quite small circles, where the feathers leave the body and increasing in size as they approach the tip. When the tail is opened into a fan the effect is spectacular.

Chapter 16

Nothing New

There is an old saying "There is nothing new under the sun"-well it is true. Here are a few examples.

No.1. At present the "buzz" word amongst progressive dairy and beef farmers is "diet feeding". This involves adding grains to the silage before feeding it to the cattle. Well, as I said my first job was in Lucan, Co. Dublin in 1959. We mixed beet pulp, brewer's grains and malt screenings.
The only problem in 1959 – we had to mix it all with a shovel. It was then filled into bags and poured on top of the silage.

No.2. We hear a lot nowadays, particularly from America and Canada about keeping the cows indoors all the time and bringing the grass/silage to them. In 1959 in Lucan, due to crop rotation the good grass for the cows was over 1 mile away. There was no way we could walk 65/70 cows along one mile of the Dublin to Cork road even in those days. Instead the grass was cut in the morning and brought to the cows. It was called "Zero Grazing". At that time the average milk yield per acre of grass in Ireland was approx. 350 gallons. With "Zero Grazing" we got 500 gallons per acre.

No.3. In the 1980's we got big bale silage. Rewind to1959 in Lucan we made plastic bag silage. It was different to the present day in so much as there was about 20 tonne in our bags and we needed the vacuum from a milking machine to extract the air.

No.4. Feeding of pigs outdoor. When I went to Gurteen Agricultural College in 1957 all of the sows were outdoors. They were penned 3 or 4 to a pen and they farrowed down in Nissan Huts with about two feet depth of straw. I thought it strange in winter to see little pigs only a few days old running around in the snow. The little pigs remained outside until they were weaned at 6/8 weeks of age. Then the pigs were put indoors and the sows went back to the dry sow pen where they were in batches of twenty with a boar to each pen. Also in the late sixties I had special harnesses made so that I could tether my sows out during the day. (see Mrs Crean's garden).

No.5. Low Wages. Farm labourers are badly paid today – what's new! In 1959 we started work at 5.30/5.45 each morning, worked till 9 o clock before we got our breakfast and finished in the evening at 6 pm. If we had to sit up at night with a cow calving this was considered part of our normal days work (no overtime). For this I was paid £5.5shillings per week. When my keep was stopped out of this and my insurance card stamped, I got the princely sum of Three Pounds one shilling and three pence into my hand. This converts into €3.99 for a six-day week. If we had to work late in the evening at hay or harvest time we got 2 shillings per hour after 6 o clock, (approx. 13 cent per hour). Is it any wonder I am a millionaire today (I wish)!

No.6. Take all the excitement and interest in the wind turbines recently built at Gartenane. They are there to generate electricity. Well, back in the forties and fifties a lot of the larger houses in the area had their own wind chargers producing electricity for that house.

No.7. At present wood shavings are very popular as bedding for dogs, hens, pigs etc. It is a highly organised business. Fifty years ago sawdust and shavings were also much in vogue. The only difference was you just went to your local friendly sawmill and got if for free. The sawmills were glad to get rid of it, you only had to fill your own. The main sawmills in the area were Sherriffs, Corries and Parrs. (See also Chapter Graham's Yard).

No.8. Electricity from water. As well as wind for electricity we also used water. At Corries there was a wall built across the river to force the water into the "mill race". The water in the mill race turned the Mill wheel, which in turn drove the turbines, which powered the forge, cornmill etc.

No.9. Swimming Pool. There was great excitement around Bailieborough five or six years ago when the swimming pool was opened. Well, 60 years ago we had a swimming pool in the area. The building of the wall or dam across the river (see above) meant there was quite a depth of water on the upper side. The kids came from miles around to swim in "Corries bathing hole". There are a lot of people around Bailieborough today in their 50's, 60's and 70's who learnt to swim there.

No.10. Walkmans. Young people today and the not so young as well, think they are great walking about with their "walkman" plugged into their ear. When I was in boarding school in the 50's we did the same thing with our crystal sets and ear phones. I did not know how they worked then and I still don't know. Whats new, I don't understand how the present day transistor radio or walkman works either.

No.11. Milk recording. In 1959 (Lucan) a regular monthly visitor was the recorder from D.D.M.B. (Dublin District Milk Board) to carry out the milk recording. Little did I think then that thirty odd years later I would be a recorder for D.D.M.B., later to become P.G. (Progressive Genetics).

Chapter 17

Alternative Enterprise

In the seventies we heard a lot of talk about alternative enterprises. This was either in remote areas where there was little or no off farm employment, or as a means of allowing an extra son or daughter to remain at home. Various forms of alternatives have been tried over the years: Tourism, Farm Relief Services, Mushrooms, Deer, Ostrich, Snail and Forestry. Some of these have stood the test of time, others fizzled out after a short period.

TOURISM

Down through the years this has taken many forms, and has been the most widespread and successful of all the alternatives. In the early days it was Farm House Holidays. With this system guests either from Ireland or Abroad could stay on normal working farms. The children could see the cows been milked and could help to feed the calves, pigs, hens, etc.

However as Ireland became more affluent, people began to go abroad for "Sun Holidays". In order to compete, farm holidays had to become more sophisticated. We saw the arrival of the Specialist Holidays. These could be Pony Trekking, Hill Walking, Painting or even Cookery.

Then we got special caravan parks, these were a lot different from the forties and fifties. In those far off days I often went on caravan holidays with my uncle, aunt and two cousins, this was in a caravan made by my uncle. If we were going to County Donegal we just set off without booking anything. If we saw a field that we liked we pulled in, found the farmer who owned it and asked if we could camp there. We usually bought our milk, eggs, vegetables and even homemade butter from the farmer. After a day or two we moved on, doing the same thing when we found another field we liked. I should point out that we had to bring our own toilet facilities with us.

FARM RELIEF SERVICES (also known as F.R.S.)

This is an employment agency set up to allow farmers to help farmers. Let me explain, very few farmers can afford nowadays to employ a workman full time. However they may need extra help at holiday time or at silage. On the other side there are farmers who do not have enough work to keep them fully employed. F.R.S. is there to bring the two sides together. Farmer A registers with F.R.S that he is available for work. When Farmer B rings F.R.S. looking for a man to help at silage, Farmer A is sent along. I myself got on average two to three days work per week for a number of years. As well as ordinary labouring work F.R.S. also trained some of their operatives in specialist work like hoof paring, fence erecting and pregnancy scanning in cows and sheep.

MUSHROOMS

Mushroom production in this area started in the late seventies but it did not really take off until the early eighties. The real impetus toward mushroom production was provided by a schoolteacher, Ronnie Wilson when he set up Monaghan Mushrooms Ltd. The concept was to use up the by products of two other farming enterprises namely poultry production and wheat growing. I should point out that County Monaghan had then and still has today one of the highest concentration of poultry production in the Country, with a corresponding mountain of poultry litter to be disposed of. County Meath had a large acreage of wheat grown every year. At that time there was very little demand for wheaten straw so it had to be either burnt, or ploughed back in. It was Ronnie Wilson's idea to use these two products which were in surplus to make mushroom compost. Prior to Ronnie's arrival on the scene mushroom compost was for the most part produced from horse manure. As horses were gradually replaced by tractors there was less raw material available for mushroom compost. Monaghan Mushrooms started a new concept of satellite farms. The company sold Plastic Growing Tunnels, Heating Systems, Cold Stores etc., to farms all over Counties Monaghan and Cavan. I myself was one of these satellite farms for a number of years. We signed a two-way contract with Monaghan Mushrooms, they contracted to supply us with compost and to buy back our mushrooms, we contracted to buy their compost and to sell back all our mushrooms.

In the beginning all mushrooms were going for processing, either bottling or canning. Under this regime we had "farm gate grading", this meant we knew exactly how much we were going to get for our mushrooms before they left our premises.

After a few years of the above system things changed, processing was phased out and instead the mushrooms went to the "Fresh Market" in the U.K. Under this system our mushrooms were not graded till they reached Monaghan and we did not know for a day or two how much we were going to get paid. Mushroom growing was very labour intensive. In our first year of production we picked our first mushrooms in late February. We then picked every day for the rest of the year including St Patrick' Day and Christmas Day. In subsequent years we had more experience and were able to avoid picking on Christmas day. We did this by picking on Christmas Eve between 8pm and 11pm and then back again on St. Stephens Day. People speak of something "mushrooming over night", this was virtually true with mushrooms. You could look into a house on a Monday and see nothing and by Wednesday the place would be covered with white mushrooms.

Since the original satellite systems concept was developed in Monaghan, systems have been set up in Counties Wexford and Roscommon. Mushroom growing is still quite active in Ireland.

DEER FARMING

There have been herds of deer in Ireland for hundreds of years. These were mainly on the estates of the Gentry and were used for Shooting Weekends etc. In the seventies the first commercial Deer Farms were set up. There were a number of deer farms set up in County Cavan but the concentration was more down South. Deer are very highly-strung and excitable. This necessitates the erection of a high rigid fence around the deer farm, otherwise the deer could not be kept at home. Another way that this excitability causes a problem, is when it comes time to transport the deer to another farm or to the Abattoir. With cattle, sheep, pigs, horses etc. you just brought out a trailer or lorry, loaded up and got on your way. If you did this with Deer they would probably be dead from a heart attack before you were one mile down the road. You had to spend about a week training the Deer. For the first day you just shut the deer in a yard with a trailer. Next day you drove them up on the trailer but did not shut them on it rather you let them come back down of their own accord, third day up on trailer shut in then later for a drive about one mile down the road, back home and let off the trailer, sixth day same story but you go a bit further, next day you are ready to proceed on your journey. There are still a few Deer Farmers operating in this area.

OSTRICH

In the eighties we saw the introduction of Ostrich into Ireland from Australia. When first introduced to Ireland we were told that there was a bonanza in Ostrich. Firstly, the flesh we were told tastes like steak, but without the cholesterol, secondly the feathers would be highly sought after for making ladies hats, thirdly the hide of the ostrich would make high class leather for ladies shoes and bags. In the initial stages there was to be big money producing eggs to hatch to supply breeding stock to further breeders. The early breeders were told to expect up to 60 eggs per female per year. These were supposed to make £500.00 each. In reality they laid about 20 eggs each and for the lucky ones the price was more like £250 - £300 per egg. I should say at this stage that an Ostrich egg is similar in size and shape to a Goose egg, rather like a small Rugby ball.
I don't know the reasons but Ostrich Farming seems to have "gone down the swanee "at this stage. It would appear that the only people to make money from ostrich were those in at the very beginning.

SNAILS

This is one alternative enterprise that never got off the ground. In the early nineties there were a number of meetings in this area organised by the County Enterprise Board, to see if there was any interest in starting up Snail Farming. I attended one of these meetings and at the end of the meeting was appointed onto a committee to examine the feasibilities. We held a short meeting at the close of the public meeting and decided that our first priority was to get some cash to fund this. I said I would get sponsorship. This public meeting was on a Wednesday night and we arranged to meet again the following Tuesday night.

Between the two meetings I contacted two companies that I had dealings with and got promises of £550 sponsorship. If I had had a few more days to work on it I probably could have doubled that figure. At the outset of the meeting on the Tuesday night it was announced by the Chairman that it had been decided that each member of the committee should contribute £50 to start the ball rolling. I said I was not in a position to pay this £50 being financially embarrassed at the time. However I said I would work for the committee as secretary or any other post that I was appointed to and that in a very short time I would have contributed a lot more than £50 in kind. I also informed the meeting of the sponsorship I had been promised.

In spite of this I was asked by the Chairman to leave the meeting, I said "I had no problem with this but if I left my sponsorship left with me". The committee met once more after that, to the best of my knowledge. That was the last that was ever heard of snail farming in this area.

ALPACAS

In the year 2000 a new type of livestock was introduced to Ireland, namely ALPACAS.

They are very much like LLAMAS, and were brought to Ireland from Australia. Their wool is much sought after. Alpacas can be white, cream, brown, black or black and white. Similar to sheep they are usually clipped or shorn once per year. Alpacas are normally very quiet and placid animals but can be ferocious when the need arises. This latter trait is used as a selling point by breeders. They suggest running one or two Alpacas with a flock of sheep to protect the sheep from foxes or marauding dogs.

Alpacas are known for their longevity, often living for 25 years or more. During a visit to New Zealand a couple of years ago I got talking to an Alpacas breeder. He told me that their main area of sales was as "Walking Lawn Mowers". At that time a lot of people in New Zealand, when they retired, moved from the cities and bought property out in the country with a half to one acre of ground with it. To keep the grass in order and also as a status symbol a lot of people were buying one or two Alpacas. Due to aforementioned longevity it would probably only be necessary to change stock once in the life time of owner.

FORESTRY

This probably has the most secure long-term prospects of all the Alternative Enterprises. It has the approval of not only the Irish Government but also the European Union. Since the late 1970's or early 1980's the area of trees being planted has been on the increase every year, most of this planting has been done on marginal land that would not grow much of anything else.

Ireland has the lowest percentage of its total landmass under timber of all European Countries. In order to redress this imbalance the Government is prepared to give farmers a grant towards the cost of planting and establishing a forest. They also promise to pay a subsidy every year for twenty years. At the end of the twenty years or shortly thereafter, the forest will commence to produce its own income.

Chapter 18

Sheep

For anyone not involved with sheep it would appear as if there has been little or no change in this sector. However, this is far from the case. I said in an earlier chapter how in the forties and fifties most farms in this area would have kept a few sheep. In most cases twenty or twenty-five ewes would have been a large flock. These would have been mainly Galway, Cheviot, Border Leicester, or to a lesser extent Suffolk. Let me say straight away that all of these breeds have one thing in common, namely their wool is white. Apart from body size, the main difference between any of the breeds is the shape and colour of the head, and therefore looking at a field of sheep from a distance, they all look basically the same. Compare this to the situation with cattle, where each breed has a distinctive colour, (see earlier chapter, Hereford – Brown, Aberdeen Angus – Black, Charolais – White, Friesian - Black and White. In this situation one can see at a glance the predominant breed of the cattle.

Back to our sheep in the forties and fifties, they were normally mated to lamb, (or yean, to give it its proper title) at the end of March, or early April. With the onslaught of the fox or disease you were doing well if you finished up with 8 lambs to sell from 10 ewes mated.

In the late 1970's things began to change. The major change was brought about with the introduction of the Ewe Premium. This was a subsidy paid by the Government on all ewes kept. In the beginning all that was required to apply for the ewe premium was that the farmer must keep at least ten ewes. This had two affects; firstly those with three or four ewes, for the most part, got out of the sheep altogether, while those with seven or eight ewes were encouraged to increase to ten or more to get the Ewe Premium. It also put some confidence into sheep farming. I never knew why but for years the sheep sector had been the Cinderella of Irish

Agriculture. An example of this exclusion was in the ruling concerning the selection of breeding stock. In cattle breeding if a farmer used a bull other than a Licensed (Pedigree Bull) that farmer was liable to prosecution and subsequent fine. Same story with pigs - the boar had to be inspected and licensed by the Department of Agriculture. Also horse breeding - the stallion had to be registered. However come back to Sheep, no such restrictions, at least not that I am aware of.

As I said, the start of the Ewe Premium put confidence into the sheep industry. Flock size began to increase. By the 1980's flocks of 500 to 1000 ewes were to be found and a few years later flocks of 1500 could be found. Also special sheep housing was erected to accommodate these extra ewes during the winter months.

New breeds were introduced; this was either to improve numbers born, or to increase lamb size. One of the first foreign breeds to be imported was the Finnish Landrace. These sheep are very different to all other sheep in so much as they have four teats or spins (all other sheep have only two). I should have said that Finnish Landrace should not be confused with Swedish Landrace, which are a breed of pigs. However, things did not work out for the Finnish Landrace, I have not heard of them in 25-30 years.

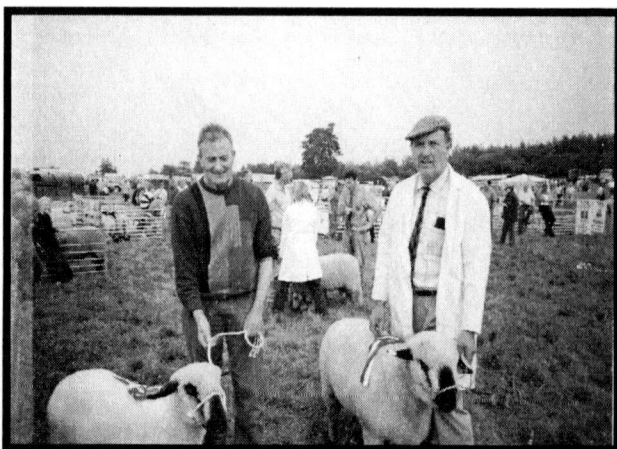

My friend John and I at Trim show

Another early arrival was the Hampshire Downs. These are renowned for early lambing, December – February and early maturity. I myself got involved with Hampshire Downs in 2002. They are affectionately known as "Hamps" among breeders.

Numerically the most successful import at present would be the Texel. Other imports include Charollais (the only similarity to the cattle of the same name is the colour white), Vendeen, Ile De France and Jacob. The latter, Jacob is the odd one out. While all the others have white wool, the Jacob is normally black and white more like a goat. The Jacob can have one, two or three horns on their head.

Jacob Sheep

Another way that sheep farming has changed is the need for dipping every year as a defence against sheep scab. All sheep had to be dipped at least once a year, usually in September. This was "Total Immersion". For small flock owners who did not have their own dip tank, there was the mobile tank provided by the County Council. It moved around the County and set up on a Lay-by or side road. The farmer was notified that the dipping tank would be in the area on Friday or Saturday and the farmer then arranged to be there on the day. This caused a lot of traffic congestion as most farmers walked their sheep to the tank. Sometimes these tanks were not emptied as frequently as they should have been, with the result that the dip liquid got very contaminated. If your sheep were not scratching before going to the tank they often were when they came back. There was a charge for getting your sheep dipped and you got a certificate stating how many that you had dipped. When speaking earlier about the Ewe Premium, I should have said that one of the conditions of application was the presentation of the Dipping Cert, No Cert -No Premium. Now dipping is no longer compulsory. This is mainly due to Sheep Scab now being less of a problem and also because there are newer chemical ways of dealing with it.

Another major change is clipping. The clipping has not changed but the cost has. When I first started Sheep in 1989 the charge for having your sheep sheared was 50p -60p per head. Wool was fetching 40p-50p per pound with an average of 5lbs per fleece, one made a tidy sum. Fast forward to 2006, cost of shearing €1.50 - €2.00, price of wool 44 cents per kilo. It does not take a mathematical genius to calculate that things are not working out right.

As far as I can see, another change is that sheep don't seem to go on their back as often as they used to do. It was a common sight years ago to see a sheep on her back, especially in springtime when they were heavily pregnant. If the feet were in the air and she was kicking she was still alive, if she was not kicking she was probably dead. They would only survive for a matter of hours on their back. Quite often while driving in the countryside, I had to stop the car and jump into a field to rescue a sheep. I might never know who owned the sheep, but that did not matter.

Chapter 19

Goats

This is another type of Livestock that has gone through a change. In the 1940's and 1950's and even the 1960's the goat was the "poor mans cow". Practically every labourer's cottage plot had its goat. At that time the "Common or Garden Variety" would normally produce ½ to 1 gallon of milk per day. In the sixties new and fancy breeds were imported. These new breeds could give as much as 4 or 5 gallons daily.

In the late sixties once again the "Health Kick" came into play. Goat's milk was supposed to be lower in cholesterol. At first the Supermarkets began to sell cartons of goat's milk. This could be either fresh or frozen. Later we got goat's cheese and goat's yogurt. Fast forward to 2006, not far from home (Bailieborough) there has been a 1000 goat unit set up to make cheese.

Goat's milk also has medicinal properties; it can relieve the symptoms of asthma, eczema and other skin complaints. In the early seventies, one of my daughters was diagnosed with lactose intolerance (unable to digest cows milk). I bought a goat and her problem soon began to improve. In fact I bought two goats, one to keep the other company. We milked one for the house and fed a calf on the other. At first we had to stand the goat up on a bale of hay so that the calf could reach. However after a few weeks, the calf learned to get down on its knees in order to get its meals.

Back to goat's milk and its medicinal properties, some of our neighbours got to know that we had goats. I started getting enquiries for goat's milk for a child with a skin complaint or an adult with a bad chest. They were usually full of praise for the benefits of goat's milk. However as soon as I asked for payment, they either got annoyed, or they suddenly had a miraculous cure and did not require any more milk. Goat's milk at that time was selling for about twice the price of cow's milk. I only requested the same price as cows milk explaining that every pint of goat's milk I supplied to someone else was another pint of cow's milk I had to buy. It did not make any difference; they were just not prepared to pay regardless of the benefits to the patient.

Goats also had a veterinary use. In the 1940's, '50's and into the 1960's, there was a disease or condition very prevalent among cattle known as "Red Water". It involved internal bleeding and was caused by a tick. This tick lived on gorse or whins and other small bushes. Now enter the goat, goats love to eat bushes and by so doing they helped to lower the tick population. I should have said that red water could be fatal if not seen to in time because of the bleeding. The veterinary surgeon normally had to administer a blood transfusion. A blood transfusion in a human is relatively simple with blood typing. However, in cattle it was a bit more complicated, one had to try to get blood from a direct relative, mother to daughter, sister-to-sister etc. Thankfully one rarely hears of red water nowadays. There are two main reasons for this, firstly with improvement in land management, there is less Gorse etc. to harbour the tick, and secondly in the sixties there was a vaccine developed.

One does not often associate goats with music but there is a connection – River Dance. When River Dance became world famous there was a renewed interest in the Bòdhràn, which in turn has made goat rearing very profitable in certain parts of Ireland, the reason being that goat skins make the best Bòdhràns.

Chapter 20

Pigs

As I have already said in an earlier chapter, in the 1940's and 50's, every farmer kept a couple of sows. These were kept to eat surplus potatoes and household scraps. Anyone sending milk to the creamery also brought back skim milk to feed to the pigs or calves.

People in town (see earlier chapter) also fed pigs. These were fed mostly on household scraps, I myself used also get skim milk from some of my farmer friends as they passed through town on their way home from the creamery.

Pigs that were fed at the back of public houses were usually very quiet during the day. There was a simple reason for this, because they were fed on slops of Guinness and other drinks they usually spent most of the day between feeds, asleep, hence the expression "Snoring like a sow".

People nowadays will get it hard to believe, but in the forties and fifties there were hundreds of pigs fed on the back streets of Dublin City. It was a common sight to see ponies and carts with two or three steel barrels going round the hotels and restaurants collecting swill . I have no first hand knowledge of it but I presume the same thing happened in other towns and cities around Ireland.

However, in the late fifties early sixties, all this came to a halt. The Department of agriculture decided that the risk of starting a "Foot and Mouth" outbreak was very high, if there happened to be imported meat included in the swill. To continue to feed pigs on swill one had to get a licence. In order to get a license you had to have facilities to boil the swill before feeding it.

Speaking of unusual methods of feeding pigs,in the late sixties I knew of a member of the Aristocracy in County Galway who used to feed hens to his pigs. His Lordship kept a large number of hens in battery cages. Some of these hens would die from time to time and had to be disposed of. They came up with the idea of boiling the carcase, feathers, intestines and all. This "soup" was then fed to the pigs. They seemed to thrive on this diet. This actually makes sense as the feathers are very high in protein. So much so, that around this same time a chicken processor, who was also a feed compounder, dried feathers, ground them and included this as a protein source in one of their rations.

In the 1940's and 50's there was basically one breed of pig in Ireland namely Large White. There had been various breeds of spotted pigs around but the Department of Agriculture decided that it was not in Ireland's best interest to encourage them, In fact anyone found breeding these "COLOURED" pigs were liable to prosecution and a fine.

At the end of the 1950's the Landrace breed of pigs were imported into Britain and Northern Ireland. Once again the Minister for Agriculture here decided that the Landrace were not for us and so banned their importation. After a few years, having seen these pigs and their progeny in Northern Ireland, a friend of mine now deceased, decided that he would risk buying a Landrace boar in Northern Ireland. This would have to be smuggled across the border under the cover of darkness. All went well, but on the homeward journey, just after crossing the border he thought the trailer felt very light behind the car. When he pulled up he discovered that the tail board had fallen off the trailer and his lovely Landrace boar was gone. He searched for hours but to no avail, no sign of his pig. Now the problem was that because of the importation ban he could not report his loss to the Police. However two days later he heard that a Landrace boar had been seen wandering around on the Northern side of the border. When he went up he found his pig in a hay shed, asleep, in a farmyard in Northern Ireland. He had to explain his problem to the farmer and beg permission to leave the pig there until after dark. Eventually the first Landrace Boar in County Cavan duly arrived at Pottle, Bailieborough.

To finish with the Landrace Breed, about two years later, we had a General Election here, and a different party got into power. The new Minister for Agriculture proceeded to lift the ban and allow Landrace Pigs to be imported freely and legally. My friend then established a Pedigree Herd of Landrace Pigs. Within a few years Landrace became the most popular breed of pigs in Ireland, either as a pure breed or crossed with the Large White.

I said at the start of the chapter that practically all farmers in this area kept one or two sows, most of the pigs produced were sold as "Suckers" or "Bonham's" (see chapter on Fair day). A couple of pigs were fed to bacon weight and killed at home for their own use This bacon was usually hung on hooks from the rafters in the kitchen. When the housewife wished to cook some bacon, either to fry it with, perhaps, turnips or to serve it boiled with cabbage, all she had to do was cut off the size of meat she needed. I should point out that there were no fridges or freezers (no electricity) in those days. Sometimes in the summer the last of the pig would get a bit smelly before it was finished.

There was no commercial fattening of pigs as such, however once again with the "Rural Electrical Scheme", all this began to change. Farmers started to build "Pig Fattening Houses". I told in an earlier chapter how in the mid 1960's I worked for a National Animal Feed Compounder. One of my duties involved calling with farmers advising them on the building of these fattening houses. I helped them to decide on site, type of building, size etc.

I remember on one occasion I was asked to call on a farmer living between Bailieborough and Cavan. I met the farmer in town and arranged to visit him next day. As there was snow about at the time, he suggested that I leave my car at the main road and that he would send the "young fellow" down with the tractor to bring me into the farmyard. Next morning I arrived at the meeting place to find the "young fellow", waiting for me, the only thing was he was about 50-55 years of age. The Father at over 80 still held the reins and made all the major decisions on the farm. This was quite common in rural Ireland at the time. It explains why there are so many bachelors around at present in their eighties and nineties. By the time their fathers died the sons were too old to go looking for wives. Fortunately the situation has been changing since the eighties. The fact that the Farm Early Retirement Scheme was introduced by the European Union has greatly helped.

However, back to the fattening houses I helped to design. These were usually for 100-120 pigs. At first these had solid floors in the dung passage; this had to be swept out every day. However, a few years later the slatted dung channel had arrived. In fact slats were introduced into pig fattening houses a few years before the same concept was thought of in cattle sheds.

Speaking of toilet arrangements, pigs have the name of being very dirty animals. This probably comes from the fact that in hot weather in Summer, sows love to roll around in mud to cool down. Now elephants also like to cool off in mud, but nobody calls them dirty.
Back to our pig-fattening house, in fact pigs are the cleanest of all farmyard animals. If their living quarters are divided into sleeping, eating and toilet areas they will use accordingly. A pig that goes to the toilet in the sleeping or eating area is usually sick. This same discipline does not apply to any of the other animals.

In the early 1960's, we saw the first of the Co-operative Pig Fattening units being set up around the Country. These were originally set up in parts of the Country where there had been no traditional pig industry. The initial size of these was for 1500-2000 pigs. A corresponding number of sows had to be placed in the locality to keep these units supplied.

Meanwhile back in County Cavan things were also moving forwards. By the seventies 500 sow units were starting to appear. In some cases these units took all pigs to bacon weight, others sold them as weaners or suckers to larger units for finishing. At present there is one group or consortium that between them has approximately 40,000 sows taking all pigs to bacon weight. Two of their units are in the Bailieborough area.

An interesting aside –in the 1960's, 70's and 80's all agriculture shows in the Country would have its pig classes. Today 2006, the only show I know of that still holds a pig class is Balmoral in Belfast, Northern Ireland.

Back in the 1950's, there was an old saying "Never eat pork in a month without an R" namely May June July and August. The reason was that these Months were the hottest. With no electricity we did not have fridges or freezers to preserve the meat.

I mentioned in an earlier chapter about "sitting" with a sow while she was farrowing. This was to ensure that she did not lie on the young pigs or eat them either. However the development of the "Farrowing Crate", was a big help in this direction. A farrowing crate was made from either wood or iron. It constrained the sow and insured that she would lie down slowly, thus allowing the young pigs to scurry out of the way. A more elaborate crate had an enclosed box at either side, in which the young pigs could sleep. The heat generated by the sow kept these boxes warm. If the boxes had not been covered this heat would have been lost in the open shed. This heat was very important for the young pigs in the first few days of their lives.

Speaking of sows eating their young, reminds me of this story. A young lad from the city came to visit his Grandfather on a farm. One night the Grandfather had to attend a very important creamery meeting. He was reluctant to go as there was a sow about to farrow. However, the young lad convinced the old man that he could look after the sow. When the Grandfather returned, he was surprised to find that there were no little pigs running around. Just then a pig was born and after a few seconds this little pig wandered up to the sows head. With one snap she swallowed it. The Grandfather roared at the young lad, who calmly said "Don't worry, that little fellow is running in and out all night!!"

Chapter 21

Horses

Of all farm animals, horses have probably gone through the biggest change. In the forties and fifties every farmer had at least one working horse, larger farms could have three or four horses. As I said in earlier chapters, all the farm work, such as ploughing, sowing, haymaking was done by horses. The horse and trap was also used to go to church/mass on Sunday, or to go socialising.

There were different types or breeds of horse used. For the heavier work, especially for industrial work most of the horses used were either Shires or Clydesdales. Both of these breeds have a lot of long hair on their lower legs, especially the back legs. This is referred to as "feathering". However, the most numerous breed in Ireland, down through the years has always been an indigenous breed, The Registered Irish Draft. As the name would suggest this is a strong working horse. It is also light enough to be used for carriage work or riding. I came across a statistic recently showing that at the turn of the twentieth century there were 300,000 registered Irish Draught Mares in Ireland. By the twenty first century there were barely 2000. Those that are left today would be almost entirely used for Sport in its various forms. As Show Jumpers they are famous all over the world. They are also used for Three Day Eventing, and for pleasure riding. Nowadays the Irish Draught is often crossed with the English Thoroughbred to produce the Irish Half Bred. These animals have the jumping ability, surefootedness and docility of the R.I. D (Registered Irish Draft) but have the speed of the Thoroughbred. A few years ago the R. I .D Breed got a great boost. This was when the Dublin Mounted Police Corps was formed. It is the policy of this force to always use R I D.

Going back to the 40's, 50's and even into the 60's in Dublin and other large cities, horses were very much in evidence. The breweries and distilleries made all their deliveries with horse and dray. The bakeries delivered their bread by horse and cart as did the dairies with their milk. When goods arrived by Goods Train into Dublin these crates were then delivered around Dublin by C I E with a horse and dray.

Right up to the mid to late sixties there were also a number of horse drawn cabs operational around the centre of the city. Of course most people by this time were using buses. In order to keep the horses and cabs on the street as a tourist attraction it was decided that some of the main hotels should sponsor these cabs. When my wife and I got married in 1964 we travelled from the church to the Hibernian Hotel in one of these sponsored cabs. The next time I saw this cab was in a Transport Museum (horse drawn and motorised) in Ashford, County Wicklow.

Travelling in style (Wedding Day 1964)

I mentioned in an earlier chapter about the horse and carts going round the city collecting "swill" from the hotels for pig feeding. We also had the "rag and bone men" driving around Dublin collecting old furniture and other "junk".

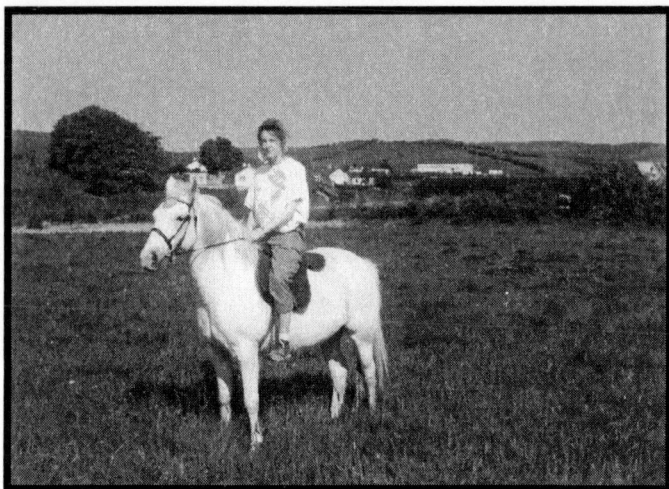

One of my daughters on her beloved "Holly" official name Holiday Grey, see other daughter on front cover.

We have an indigenous breed of pony in Ireland, namely the Connemara Pony. In the past this pony was used extensively for work around some farms, particularly in the West of Ireland where the terrain was rougher. They were always renowned for their surefootedness. Of course nowadays the Connemara's are bred almost entirely for pleasure riding, They are exported all over the world. Sometimes they are crossed with the thoroughbred to add height. One of the most renowned horses to ever Show Jump for Ireland was a little Horse called "Dundrum", ridden by Tommy Wade. Dundrum was the result of a Connemara/Thoroughbred cross. For several years in the sixties Dundrum and Tommy Wade were the anchor men on The Irish Nations Cup Team competition for the Agha Khan trophy at the R.D.S. Horse Show. I bred Connemaras for a number of years in the eighties and nineties. I should not say so myself, but I had a modicum of success at the local show over the years.

I could not finish a chapter on things equestrian without mentioning the humble Donkey. Back in the forties and fifties the donkey was the poor man's horse. They were used to bring milk to the creamery, to bring turf home from the bog and in some cases to drive to Church on Sunday. My own experience with donkeys began at the age of eleven or twelve. In the fifties we did not have a "Refuse Collection Service", it was up to every household to make their own arrangements. In the case of my own family we had an "Ash Pit". This was a walled in area at the bottom of our yard. Into this pit we put ashes, tin cans, bottles etc. This pit had to be emptied two or three times a year. When I was about eleven or twelve the job fell to me. In the beginning I borrowed a large Spanish Ass from our neighbour T.R. Smith. I had to climb into the pit and shovel the contents onto the cart, and then drive up through the town to the "Land fill site", known as the Bottles, (see earlier chapter). It was a smelly messy job but as a young lad I felt very important driving about with my donkey and cart. However a couple of years later I advanced to a pony called Bunny. (See Grahams Yard). This job was mine till I got married in the mid sixties.

Before leaving about donkeys, I would ask my readers to take a good look at a donkey if ever they get the opportunity. You will notice a strip of dark hair running right down the donkeys back and across its shoulders. This is in the shape of a cross. It can be found on every donkey, regardless of where they came from and is said to be, in commemoration of the time Our Lord rode into Jerusalem on a donkey, all those years ago.

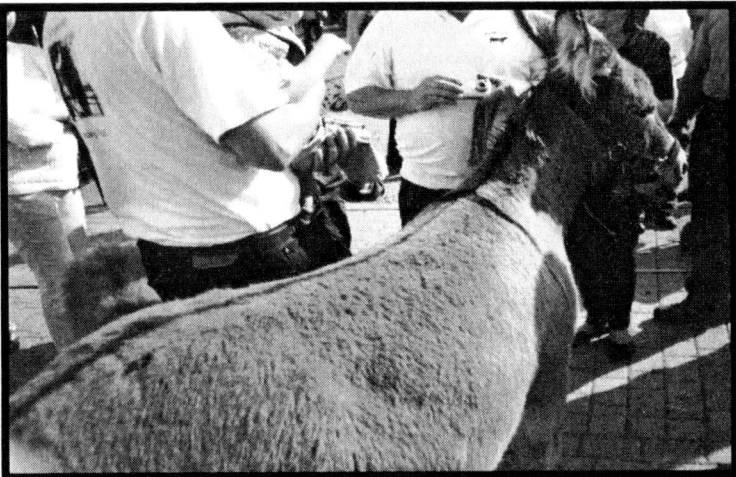

Donkey - note cross on back

Two other little snippets, if you mate a donkey mare with a pony stallion the offspring is known as a Jennet. If the mating is the other way round, donkey stallion, known as a jack ass with a pony mare the offspring is a mule.

I should point out that a pony never becomes a horse. The first time I came across this misconception was while in Gurteen Agric. College back in 1957. A fellow student from Belfast (can't remember his name only his nickname "Squeaky Mc Cauley") asked the question six months into the course. "When will the pony grow into a horse?". The pony in question was 14 or 15 years of age. On a recent trip to New Zealand I was surprised to discover that my grandson had the same idea.

All things equestrian are measured in "hands". A hand is 4 inches (100mm). This will vary from the Shetland pony at approx 9 hands = 36 inches = 90cm, to a large Draught Horse at 17 hands = 68 inches = 170 cm. All the various Pony Breed Societies have their minimum/maximum height permissible for inclusion in their Breed Register. In the case of the Connemara the minimum is 13 hands and the maximum 14.3". Most Horse Breeds start at 15 – 16 hands. There is one exception to this. There is a breed of Miniature Horse, called Folabella. It comes from South America and is smaller than a Shetland pony.

I have spoken at length about the work done by Horses, however before they start to work they have to be trained or broken, This takes weeks of patient handling, starting with getting used to having a steel bit in their mouth and then getting used to the weight of something on their back. This can be either a rider or a cart etc. This training normally starts when the horse/pony is three years old. Herein lies another story. If you had bred the animal yourself no problem, but if you had bought it you could be in trouble, because of the way the age of a horse is calculated. Let me explain- if a foal is born on March 1st 2005 come January 1st 2006 this foal is known as a yearling, come January 1st 2007 and it becomes a two year old, and so on. However lets take an extreme example – if our foal is born December 1st 2005 once again it becomes a yearling on January 1st 2006, come January 1st 2007 it becomes a two year old, January 1st 2008 three year old even though it is in reality only twenty five months, hence the problem, officially it is a three year old but at 25 months is not sufficiently mature to commence training.

In the first year after training it is only subjected to light work. A horse is five or six years old before it really comes into its own, regardless of what the work is. Provided they are looked after properly horses can continue working till they are fifteen, sixteen or even up to twenty years old.

Let us look for a little while at the harness or "Tackle" worn by the Horse. We will start with a riding horse. Firstly on his head he wears a bridle. This comprises straps of soft leather stitched together; the metal part which goes into the mouth is called the "Bit". The reins are attached to each side of this bit. These reins are the means by which one controls the Horse, pull on the left rein to turn left, right rein to go to the right and pull both reins evenly at the same time to stop. On the horses back he has a saddle on which the rider sits. The strap which goes around the horses belly to keep the saddle in place is called the Girth. If the horse tends to be a bit excitable or unruly we could fit a Martingale. This comprises of a leather strap which is attached to the Girth Strap passes between the horse front legs known as forelegs, is then divided into two strands, these straps are attached to each side of the bit. This can be adjusted to allow the horse adequate freedom to walk trot etc. However if he tries to rear up on his hind legs in order to dislodge the rider he is prevented from doing so.

Now the harness or Tack for a working or draught horse is naturally different from that for a riding horse. Firstly the harness has to be stronger in order to withstand all the pulling and carrying that the work horse is required to do. The bridle is replaced with the "Winkers". On the neck we have the Collar. This has to be well padded and fit properly because it is by means of the collar that the cart etc is actually pulled. Around the collar we attach the hames. There are leather straps, called draughts, in the case of light carts or traps, or chains called traces in the case of heavy or stiff cart or other farm machinery. These draughts or traces were hooked onto the hames at one end and onto the cart at the other end.

After the collar we have the Straddle. In some places this is called the Saddle as in a riding horse. Once again this has to be well padded and fit comfortably as it is the straddle that actually carries the weight of the cart or other machinery. In the case of a light cart or trap there are straps (leathers) coming out of the side of the straddle which have loops at the ends. The shafts of the cart/trap are slipped into these loops. This is done by backing the horse between the shafts. With a heavy or stiff cart, that is used for heavier work the system is different; here there is a chain connected to the two shafts. When the horse is backed between the shafts this chain is dropped into a 'groove' across the top of the straddle. Both the chain and the groove have to be greased regularly so that the chain can move with the swinging of the cart.

Behind the straddle we have the 'britchen'. Again this is a collection or series of leather straps which are hooked to the back of the straddle. It then lies loosely on the horses back and drops over his sides and down behind his rump. In the case of trap or driving carriage harness we also have a "crupper". What's a crupper I hear my readers say- it is a stiff ring of leather at the back of the britchen, through which you pulled the horses tail. This causes the horse to hold his tail a bit higher. This was purely for cosmetic purposes, it was considered more stylish to have the horses tail held up in this way. I said earlier that the collar does the pulling and the straddle does the carrying, the britchen acts as the brakes. It is attached to the shafts of the cart with either straps or chains. When the driver pulls on the reins to stop the horse, the horse in turn braces himself against the britchen and this stops the cart from coming forward. This is also very important if the horse and cart is going down a steep hill. If we did not have a britchen, the cart would keep coming forward and could eventually knock the horse. However, with the britchen the cart is held back. Another important piece of "tackle" is the bellyband, this is either made from leather or rope depending on whether it is used on a trap or working cart. This is attached to one shaft of the cart/trap, passes under the horses belly (hence the name) and is then attached to the other shaft. The purpose of this strap or rope is to keep the cart from tipping up. If too much weight is placed at the back of the cart it might tip up backwards taking the horse with it, the bellyband prevents this.

"The same thing from opposite sides of the world"
Top Photo: Irish work horse complete with harness Bottom Photo: Working horse in New Zealand showing freeze brand number for identification

Chapter 22

Hay Making

Nowadays with modern machinery haymaking is a simple operation, provided the weather is favourable. You can cut hay on a Monday and have it in the shed by Friday/Saturday. In my young days this could take anything from six to eight weeks.

Horse drawn mowing machine, note the breast pole has been shortened to use with a tractor

Firstly, you cut the hay. In 1950, with mowing machine pulled by two strong horses it took about half a day to mow 1 acre, 2006 with large tractor – half an hour. Turning: In 1950 – 3 or 4 men with hand forks spent most of a day turning and shaking out this acre, 2006 tractor with hay turner, ½ hour tops.

Rearing: 1950- it had to be turned by hand once per day for 3-4 days depending on weather. 2006, 3 or 4 times in one day with machinery Lapping: The whole idea of haymaking is to get the hay off the ground and dried as quickly as possible. In 1950 if the crop was heavy or the weather unsettled it had to be lapped. This was a very backbreaking job for somebody as tall as myself. It meant bending down and lifting an armful of hay, this was wrapped around your arm, in such a way that when you placed your lap on the ground, and removed your arm it looked like a little bun, but where your arm had been there was now a tunnel going right through your lap, to allow air to circulate through and so dry the hay. The hay remained in those laps for 3-4 days and then when you got a really good day they were all shaken out again. After another day the hay was then put into "hand shakes". These were small cocks of hay containing about 5-6 laps. These had to be straightened off so that the rain, if it came, would be repelled. After a couple of days these "hand shakes" were shaken out and the whole field gone through again with hand forks.

2. Cocking: Hopefully at this stage it was dry enough for cocking. To do this it had to be gathered into rows. If it was a small field this could be done by using forks and rakes. When the field was bigger there was machinery for the job (yes, even in those dark ages). The first of the machines was a horse drawn wheel rake. This had two steel shafts, just like a cart between which the horse was tackled. Behind the horse was two steel wheels about 4 foot high and about 6 feet apart. Connected to the axle you had the frame which consisted of a seat for the driver and behind the driver you had 10-12 steel teeth or tynes. When the horse moved forward across the hay these teeth gathered in the hay.

The driver would start at one side of the field, gathering the hay as he went along. When he had the full of the rake gathered he pulled a lever, this lifted the tynes, allowing the hay to drop out. When the horse moved two steps forward the driver then let the tynes down again to recommence gathering. When full again, same procedure. When he got to the other side of the field, he then turned and came back across again. Each time he came to the row of hay started on the first trip he pulled his lever, thus increasing the length of the row. When he had repeated this, back and forth over the entire field he finished up with long rows of hay running the full length of the field. These were knows as "wind rows".

My wife on wheel rake

If the day was dry enough and the weather looked good you started to build hay cocks straight away. If it was not dry enough it could be left in the "wind rows" till next day.

If ready to go ahead, these wind rows could be lifted by hand with a hay fork or we could use an implement that I have neither seen nor heard of for over 40 years – "a tumbling paddy". This was a wheel less horse drawn rake. It was made of wood, except for the traces or chains. It had no shafts just traces (chains). There was a heavy wooden frame about six foot wide, again with a number of teeth only this time they were wooden and it had two handles for the driver to steer it along by. The rake was attached to the horse by hooking the end of the traces on to the "hames" on the collar around his neck.

The object of the exercise was to gather in the wind row to a pile in the centre. To do this the driver positioned the rake at the end of the wind row facing into the row and the horse standing on the row of hay. When the horse was told to move forward the rake gathered in the hay in the same way as the wheel rake. I should have said that the driver was walking behind (no seat this time). When he thought he had enough hay gathered in he let go of the handles, gave them a slight push forward, thus causing the rake to tumble (hence the name). When it tumbled it dropped the hay. The driver then caught the handles again putting the rake back in the "upright" position ready to start raking again. When there was sufficient hay gathered in, the other men in the field started to build the cocks.

3. Building the cocks. You started with a circle of hay about 4 feet wide and then you built on this till the cock was about seven foot high. You could not throw the hay on in any old way; it had to be shook and straightened out. You also had to slope it in slightly so that it was narrower at the top than the bottom, so that the rain ran off. When your cock was built you had to "dress it down" and "pull the butt" and "head the cock".

Cocking Hay

The last thing was to "rope it". This was normally done with a hay rope which you twisted out of the "butt" of the cock. There was a tool for this job. Alternatively you could use a rope known as "hairy ned". When the entire field was cocked the last job was to "clean rake" the field. This meant going over the entire field with hand rakes to gather up any loose hay that was lying around. (See front cover)

Working in the hay field on a hot summer day was very thirsty work, and so it was necessary to have a can of liquid sitting in a cool place where it was easily accessible when required. Different liquids where favoured by different people, some liked "spring water", others water with a little milk added to it, others preferred stout. My own favourite was buttermilk about two weeks old that was beginning to "curl up and look at you". One mouthful of this would "slake" any thirst.

Speaking of refreshments, the highlight of working in hay field was "Tea" in the hay field. Sandwiches, whether ham or jam never tasted as good as when eaten while sitting on a "lap" of hay. Although vacuum flasks had been invented, tea from a "sweet can" tasted much better.

Tea in Hay Field

Now back to the hay. It has now been cut for 7-10 days. It needs to remain in the cocks, in the field for a further 4 to 6 weeks. When the time came to bring it in there were three possible methods. 1. Tracing, 2. Haycart, 3. Horse and Cart.

Tracing

If it had just to be brought from the field directly to the hay shed or haggard the simplest way was to trace it in. This involved putting a chain right around the cock, attaching this chain to a "swingle- tree" which in turn was attached to the horse. You then proceeded to pull or trace the cock in.

Haycart

If it was necessary to go onto the road, then you used a haycart. This was a low flat cart with a tipping body. You tipped the cart and then backed the horse and cart till the tip of the cart was under the cock of hay. Two ropes were then pulled out and brought round the cock and hooked together at the back. These ropes were attached to an iron bar going across the front of the cart. At either side of the bar you had a long wooden handle which was attached to a rachet. The idea was that two men, one each side, pumped their handle up and down thus winding up the ropes around the bar. As the ropes were wound up the hay cock was being pulled onto the cart. When the cock reached half way the weight of it caused the cart to drop into the flat position. You continued to pump the handles till the cock was at the front of the cart. You then proceeded to sit on the front of the cart and drive for home. By the way, an example of this type of hay cart complete with stuffed horse can be seen in front of the health centre in the village of Dunshaughlin. This is on the left hand side as you enter the village on the N3 from Dublin to Cavan.

Horse and Cart

If the journey from field to shed were any distance, it would take too long to draw the hay one cock at a time. In this case you forked and built the hay onto an ordinary horse and cart. This was the method I had most experience with. On a horse cart you could get 4 or 5 cocks depending on size and on the pony cart you could get 2 or 3 cocks. Usually on the journey home there would be three or four kids riding on top of the load of hay. Normally I was the driver, to the envy of the others. When we got to town everyone had to slide down as the load of hay could barely fit through the gateway.

Once in the yard all the hay had to be forked up onto the loft. If it was suspected that the hay was still damp, there had to be air tunnels left in the hay to allow air to circulate. To do this we filled bags with hay and placed them in different places around the loft floor. We then filled in the remainder of the floor. Then, as we filled the hay higher we proceeded to pull the hay bags up, thus forming several tunnels or chimneys throughout the loft. On a hot day the temperature on the loft was like an oven. There was a galvanised iron roof on it, which would burn you if you touched it.

We finally have the hay "saved" and in the loft approx 8 weeks after we started. Now go back to the beginning of the chapter – 2006 – 4 or 5 days.

Chapter 23

Tillage – Grains

Let me say at the outset that I am not as well versed on tillage as I am on livestock. One does not normally associate Co Cavan with tillage. Grain growing at the present time would be more associated with counties Meath, Louth and Kildare. However in the past things were different. This is obvious by the fact that there were two corn mills in Bailieborough and another one two miles outside of the town. As I said in an earlier chapter, most farmers around here in the fifties grew a couple of acres of corn. This was mainly to feed their livestock. There were two or three mobile threshers going around the locality threshing the corn every autumn.

In order to sow corn in 2006, the ground would be ploughed with a large tractor, then rotavated using another tractor, then the seed would "go down the spout" with the fertiliser in a combined seed drill. Step back to 1940/50's and it was all Horse Power. The plough was pulled by two horses and the Spring Tooth Harrow was also pulled by two horses. However the seed was sown by man power using either a Fiddle or a Sheet. Let me explain, with a Fiddle you had a bag of seed strapped to your back, the neck of this bag allowed the seed to flow down onto a round table. To this table was attached a long narrow handle. When you pushed this handle in and out it caused the table to rotate, the seed went down through a hole in the bottom of the table and the rotating movement caused the seed to scatter about two feet either side of the man as he walked across the field. Pushing the handle in and out was rather like playing the fiddle, hence the name.

With the "sheet" you tied a sheet around your neck and under one arm, rather like a sling. You then put seed in the sling. One hand was used to keep the sling in place while the other hand scattered or broadcast the seed, again as you walked along. The seed then had to be harrowed in.

Let us look at the harvesting of this corn. Similar to the hay it took a lot longer in the fifties than it does today. In 2006, with the use of a Combine Harvester the entire operation takes about one hour from the time the harvester goes in the gate until the grain is in the grain store, and the straw is baled and in the shed. Compare that to the 1950's when the process could take anything up to four months.

In these parts the corn was cut with a horse drawn mowing machine, similar to the one mentioned in an earlier chapter, for cutting hay. However, it had to be modified. There was a Sheaving Board attached and also a second seat. The first seat was for the driver, the second was for the Sheaver.

The Sheaving Board was attached behind the cutting blade and was set at an angle. The reason for this angle was to collect the corn as it was cut, rather than let it fall straight to the ground as one would with hay. When the man on the sheaving seat could see that he had enough collected, he pushed a lever at his feet, which lowered the board. Then with his Sheaving Rake he pushed the collected corn off the sheaving board. The driver then went on to cut the next sheaf. When the sheaf was dropped it was picked up by the Gleaners who straightened it out and then tied it in a sheaf with a few strands of corn. This job was normally done by women or young men, as it involved a lot of stooping. After the Gleaners came the Stookers. They gathered up five or six sheaves, stood them on the ground together, with the thick end or butt on the ground and the grain end on top. These were tied together, again with some strands of corn. When tied together like this they were known as stooks, also known as "Old Grannies".

The corn remained in these stooks for two or three weeks depending on the weather. Then they were gathered in and built into a stack. In the stack the butt was towards the outside and the grain towards the inside, placing layer upon layer. The top layer was stood up and then tied like a stook. This was to keep the rain out.

Again depending on the weather, the corn normally stayed in the stack for a further three to four weeks. It was then brought into the haggard, which was a small field or garden close to the house. This garden was used to store all the winter feed. Similar to hay, the corn could be traced in or brought on a hay cart and then built into a Pike. This Pike was built in similar fashion to the stack only bigger. In fact all of the stacks from the field were put in the one Pike. If the farmer had three fields of corn he would have three Pikes.

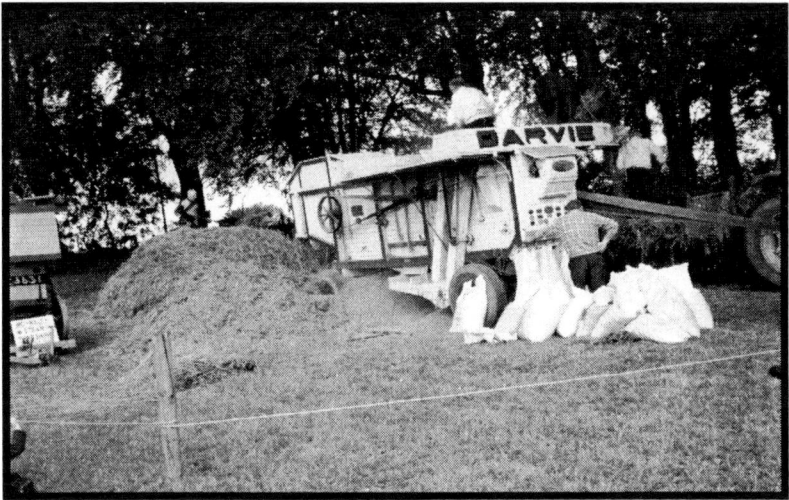

Threshing Machine

At the end of October or early November, the travelling 'threshing machine' started on it rounds. When it arrived at a farm it was pulled right into the haggard. The thrasher required about eighteen or twenty men to attend to it.

Firstly, you needed four or five men opening the pike and drawing the sheaves near to the thresher.

Two men pitching the sheaves up onto the thresher.

Two on their knees on top of the thresher, to remove the bands and feed the corn into the thresher.

Two on the ground at the back of the thresher in charge of the sacks of grain.

Two to keep the straw pulled back.

Four more taking that straw and building it into a rick.

Two with a horse and cart transporting the sacks of corn to the granary and another in the granary empting the sacks.
You also needed a couple of stout "Gossoons" running around with sticks and shovels killing the rats and mice that were inevitable at any thrashing.

I should have said at the outset, that the thresher was powered by a large tractor. There was a pulley block on the tractor and another on the thresher. A large belt ran from one block to the other. When the P.T.O. (power take off) of the tractor was put in gear this started the pulley block to spin round, which in turn caused the belt to revolve, which turned the pulley block on the thresher causing the thresher to start up. It was important to have the correct amount of tension on the belt, too loose and the belt fell off, too tight and the belt broke. For this reason it was important to have a good man in charge of the tractor.
When I spoke earlier about all the help required for the thresher, I should have said that the neighbours all gathered round – you come to me today and I go to you tomorrow. The farmer's wife had to feed all these hungry men. A normal thresher dinner would be home cured bacon and cabbage, lots of potatoes with home made butter and buttermilk to wash it all down. This was fine the first day, but if you followed the thresher for a week it could get a bit monotonous.

The scene I have just described was common in these parts until the mid forties. However, science was marching on; by the late forties the Reaper and Binder, had been developed. It too was pulled by two horses, or sometimes three for the larger models. As the name implies it cut the corn and also tied the sheaves. This did away with the back breaking job of having to stoop to pick up the corn to tie into a sheaf. However, everything else remained the same, the corn still had to be stooked, stacked, piked and eventually thrashed.

The first Reaper and Binder that I knew of in this area was owned by cousins Ronnie and Fred Clarke. By then it was pulled by a tractor. This would have been in the mid fifties. By the early sixties these same cousins had bought their first Combine Harvester. Unlike the present day Combines which are self propelled, this one had to be pulled by a tractor. It also differed in its method of collecting the grain. The modern Combine retains the grain in a bulk tank, which can hold approximately half a tonne of grain, and this grain can be transferred into a bulk trailer, even as it drives along. The original Combine filled the grain by two spouts into two jute sacks which could hold sixteen stone (100kg) each. There was a man standing on a platform to change these bags. When full they were tied and then let slide down a chute to the ground, from there they were loaded onto a tractor and trailer for transfer to the granary.

As I said, in the fifties and even into the sixties there was not much corn grown in these parts. Fred Clarke would have been one of the bigger growers. As well as sowing his own land, he also rented a lot of ground as well. However, farmers were reluctant to rent ground for tillage so Fred came up with what was a novel idea at the time, Share Cropping . We hear a lot of talk nowadays about share milking in Australia and New Zealand. Back in the mid seventies Fred was very much ahead of his time. He made an agreement with the farmers whereby they supplied the land and Fred supplied the machinery, seed, fertilizer, sprays and labour. At harvest time the crop was split 50/50. This continued until about 1980. Following two or three very wet seasons, Fred had grave difficulty in getting the crop harvested. When he did get it cut the yield of grain was much lower than in normal years. The moisture content was also very high, 30%-35%, compared to a dry year, at 16%-18%. As oil prices had started to rocket, the cost of drying the grain was prohibitive.

The growing of barley or oats in this area has never taken off since. In fact what little was being grown in the 50's and 60's no longer exists. At the present time, you would have to travel most of County Cavan in order to find a total of ten acres of grain. The exception to this however, would be in the last five or six years where some farmers are growing barley and even maize to use in whole crop silage.

To summarise, grain growing in 1940/50 took approx 3 to 4 months from cutting to granary. However when it did get to the granary it was usually dry enough to store over the winter. In 2006, from cutting to granary is about ¾ hours but this grain could come off the combine at 20%-22% moisture and then have to be dried artificially.

Chapter 24

Tillage (Roots)

The main root crop in the 1950's was potatoes. As I said in an earlier chapter most farmers grew a couple of acres of potatoes. When the ground had been ploughed and harrowed the drills were opened with a horse drawn drill plough, then came the F.Y.M. This was spread in the furrows by hand using a four- pronged fork known as a manure grape. Next we need the seed potatoes. These could either be round or cut. With round seed you planted a complete potato about the size of a small hen egg. For cut seed you took a larger potato and cut it into pieces, making sure to have a least one but preferably two eyes per section. It is from the eyes that the sprout starts to grow. When using cut seed it was important to dip the cut side in ashes or some similar material. This helped to keep the worms from eating the potato before the sprouts were formed. The strain of seed sown depended on the time of planting, there were earlies, semi-earlies or main crop. In these parts in the 50's a popular seed for earlies was Sharps Express. They were a kidney shaped potato. I have not heard of them for years. Another breed of potato that seems to have disappeared completely was Epicures. They were a large potato, but tended to be very wet or soapy.

In these parts in the 50's and 60's people tried to get their earlies planted before St. Patrick's Day. This would be a couple of ridges planted in the garden, rather than in the fields. Tradition also stipulated that you dug your first dinner for the Sunday before "the twelfth" (July 12th).

Back to planting our main crop potatoes. The seed dropped by hand on top of the F.Y.M. in the furrow. It was common practice for the women to carry the seed in their apron, while the men carried them in a bucket. When the seed was dropped, the drills were then closed using a horse drawn drill plough.

When the potato sprouts begin to appear above the ground they had to be covered up again. This involved putting more clay on the ridges or drills to keep the late frost from killing the young shoots or stalks. In the case of the potatoes planted in the garden, this was done with a shovel and was called shovelling. In the field situation it was done with a drill plough and was called moulding.

As the summer progressed the potato crop had to be sprayed a number of times depending on the weather to protect against blight (remember the famine). Nowadays there are several proprietary brands of spray available, but in the fifties it was home made, the main ingredient being Blue Stone. In the case of garden potatoes, the mixture would be applied using a Knap Sack Sprayer carried on one's back. For the main crop in the fields we used a Horse Drawn Sprayer. This was very simple and basic compared to the modern sophisticated Tractor Sprayers of today. Crude it might have been, but effective none the less.

As I said earlier, the Early Potatoes were ready for harvesting in mid July. These were "dug" with a spade or digging fork as required. One normally dug just enough for two or three days at a time.

The main crop would normally be ready late September or early October. Here we used horse power. Two horses walked, one on either side of the drill pulling the Digger. The wheel of the digger lifted the potatoes and fired them out scattering them in all directions. Then came the back braking job of picking the potatoes. Teenagers were best for this job. While these teenagers did not relish the job, they agreed to do it because it meant getting a few days off school!

When the potatoes were picked they had to be sorted. The small ones or chats were left for immediate feeding to pigs. Next we had the Seeds; these had to be stored very carefully for next year's crop. The bulk of the crop fell into the category of Ware or eating. These had to be further sub-divided, anything that was marked or damaged had to be put separately, either for immediate use or again for pig feeding

I said earlier that part of the potatoes were grown as a Cash Crop. Some of these were sold straight after digging, having been filled into 50 kg Jute Sacks. The remainder would be sold over the next four or five months, whenever cash was needed. They could be sold in different ways:
At home to potato merchants who went from farm to farm after digging time.
In town at the Market House on Fair Day
To shopkeepers in the town
Direct to householders.

My mother for years got her potatoes delivered by the late Tommie Heaslip, Killinkere. His workman or boy", Ned Flynn is still going strong on his own farm. Ned would deliver to several houses in town at a time. They were in 8 Stone = 1Cwt =50 Kilo Jute sacks. For years the price remained the same 8/= =40p = 50 cent for 50 kilo. Compare that with 2005 where the price is approximately 70 cent per kilo.

Potatoes that had to be stored for any length of time were put in a pit. To make a pit of potatoes you needed a level piece of ground high enough in the field so that there was no danger of flooding. Next you needed a quantity of straw or dried rushes. You placed a layer of straw/rushes on the ground, about three feet wide and twenty feet long. The length would depend on the amount of potatoes to be stored. The potatoes were gently placed on the straw and built up rather like a pyramid. When your pyramid was approximately three feet high by the length of your straw, you then placed more straw on top of the potatoes covering them to a depth of about two inches. I should have said that you had to be very careful not to include any rotting potatoes in your pit. It would be like one rotten apple in a barrel of apples.

Anyway, back to our pit. Having got a good layer of straw over the entire pit including the two ends, you then covered the whole lot with clay. The last thing to do with our pit, was to dig a small trench or water track around it to take away any rain- water that might otherwise soak into the pit and so rot the potatoes. Over the winter months whenever potatoes were needed, either for home use or for sale, you gently removed the clay and then the straw from part of the pit. When you had extracted what you needed you put the straw and clay back.

In the late fifties early sixties things started going wrong with the potato market. Some years the price was so bad that farmers could not afford to sell. Instead they started feeding more to pigs. To help with this operation we had Mobile Potato Cookers going from farm to farm (similar to the thresher). These machines could hold about ½ a ton of potatoes at a time. Firstly, the machine washed the potatoes and then cooked them by steam. When cooked they were stored in a shed to be used when required. This was known as Potato Silage. It is ironic to think that some farmers who did not start to make grass silage until the mid seventies, were making potato silage ten or fifteen years earlier.

Another important use for potatoes in the fifties and sixties was the making of Boxty Bread. This was similar to our modern day potato bread only it was made into a large loaf similar to soda bread. It was made in a "Griddle Pan" and was about two inches thick. It was very popular at Christmas time. In fact in some households it was considered more of a treat than Christmas cake. I can remember when visiting in some houses in the country, being offered a cup of tea, and the choice of Christmas cake or Boxty. It could be eaten cold with butter on it or it could be fried. Quite often country people would give a cake of Boxty to their friends in town as a Christmas present.

Potatoes were not the only root crop grown in the 50's. Farmers also grew turnips for human and animal consumption along with swedes and mangolds for the animals. These swedes and mangolds were often grown to be fed to sheep over the winter,by allowing the sheep to graze them directly from the drills in the field. Kale (a member of the cabbage family) was also grown to be fed direct to cows. For the past twenty years or more one would travel a large part of County Cavan to come across even one acre of roots. However, in the past year or two some farmers have started to grow them again, to be strip grazed by cows over the winter.

Another crop I must mention is SUGAR BEET. Let me say that in some countries sugar is produced from sugar cane, but in Ireland it is produced from sugar beet. Sugar beet is rather like an overgrown parsnip. To my knowledge it was never grown in this area. It was a very important crop in other parts of the Country.

My experience with sugar beet took place while I was in Gurteen Agricultural College in 1957-58. At that time the College grew about twenty acres of sugar beet every year. It was a very labour- intensive crop at that time. The ploughing and sowing was done by tractor but most other operations were done by hand. The beet was grown in drills, and when the little seedlings were about three inches high they had to be "thinned".
This involved tying jute sacks around our knees and then getting down and travelling along on ones knees between the drills. You had to pull every second seedling, to allow more room for the one that was left to grow.

At harvest time, again it was mostly hand work. The first job was "Pulling", you walked between two drills, caught the leaves of a beet from each drill, one in your left hand the other in your right. You then pulled up the two beet. When you had them pulled you clapped them together, to remove the surplus clay. You then placed them in a small clamp with the beet facing in and the leaves on the outside. When you had enough in your clamp you placed more loose leaves on top. The beet was left in these clamps for a few days to dry out.
The next stage was "Snagging", this involved removing the leaves. This was done with a heavy knife or cleaver. The blade was about fifteen inches long, three inches wide and very sharp. You took one beet in your left hand, the cleaver

in the right and with one swift movement you removed all the leaves. This operation took a bit of practice, if you cut off two much you were wasting sugar, but if you left bits of leaves behind the factory was not pleased.

Once snagged the beet was again put in clamps. From the clamps they were loaded onto horses and carts. To load them we used a "BEET SPRONGE" which was a special fork with about ten prongs on it. These prongs had round lumps at the end in order not to damage the beet.

The horse and cart brought the beet out to the road side where again it was placed in a clamp on the grass verge. This clamp could be twenty or thirty yards long. From here it was loaded onto lorries or tractors to be transported to the sugar factory. Back in the fifties there were very few tipping trailers so in order to get the beet out of the lorry a strong jet of water was used. This was strong enough to flush the whole lot out.

Nothing was wasted in those days. The tops could either be fed to sheep by allowing them into the field to pick them up from the ground. Alternatively, they could be loaded onto the horse and cart and brought into the farmyard to be fed to the cows.

For every ton of sugar beet that the farmer sent to the factory he was entitled to get a certain amount of beet pulp back. Unlike the beet pulp commonly used today, which is dried, in the fifties this came back wet (and hot). It had to be ensiled between two walls and covered. It could then be used as cattle feed over the rest of the winter.

In the fifties we had four sugar beet factories in the country. Gradually over the years this had been reduced to one. While this made sense to the sugar company, it made life more difficult for the farmers. It meant much longer journeys to get their beet to the factory and because of this distance, they were often unable to avail of their "wet pulp" entitlement.

However, these beet growers were not prepared for the "bombshell" which was dropped in 2005. The "powers that be" in Brussels decided that it would be more economical to import sugar from South America, or some such place. The outcome of this is, that 2005 will have been the last year sugar beet was grown in Ireland.

I said earlier that back in the fifties sugar beet growing was very labour intensive. Over intervening years the actual growing and harvesting has become much more mechanised. However, sugar beet growing still gave a lot of employment every year, between growing, transporting ant processing. It also gave a lot of employment down-stream, where the by-products – beet pulp and molasses were further processed into various forms of animal feed.

Speaking of a farming enterprise that shortly will be no more, reminds me that another industry that has long gone. I am thinking of course about flax growing. As a small boy in the forties, I often watched the men tramping the flax in " Corries Flax Hole". The purpose of this was to "cure" or "rett" the flax. This was in the field where Irish Foundries (now Bailieborough Foundries) was later built. The men normally wore nothing but bathing trunks as they waded about in the dark/green water. I seem to remember it was also quite smelly! There were a couple of other flax holes in the area. Across the road from Corries flax hole there is the remains of Kelly's flax mill or scutch mill. Funny thing to tell, I cannot remember any flax growing.

The flax industry and subsequently the linen industry are long gone, not only in Co. Cavan but also in the Republic of Ireland in general. However, the linen industry did survive in Northern Ireland right into the 1990's. I should point out that since the 1970's there was little or no flax grown in Northern Ireland either. To the best of my knowledge the fibre (hemp) was imported from Holland.

Chapter 25

Livestock Breeding

A number of changes have taken place over the past 50 odd years. In fact about the only thing that has not changed is the bit about the "Birds and the Bees"!

Cattle - The biggest changes have taken place in this section. Sixty years ago all breeding was the natural way. Needless to say not every farmer kept his own bull. Instead, in each district there was one farmer who kept a bull "for the country". As I said in an earlier chapter, these bulls had to be pedigree, inspected and licensed by the Department of Agriculture.

Twice yearly, in spring and autumn, there was a bull show held on the Market Square in Bailieborough. Any local pedigree breeder who had a young bull for sale had to present it here. If Department of Agriculture inspectors thought your young bull was of sufficient standard they put a Tattoo in the bull's ear. If on the other hand they turned down the bull, they proceeded to punch a hole in the ear, so that it could not be presented at another venue; there was no arbitration.

These young bulls could either be sold privately or they could be brought to a Bull Show and Sale. The main sales for this area were in Carrick on Shannon in early February, or the R.D.S. in Dublin two weeks later, and also a supplementary sale in Carrick on Shannon in April.

Sadly the Bull sales in the R.D.S. have been discontinued since the early nineties. However Carrick is going from strength to strength. In fact it is now a very modern sales complex, with all the stalls inside, cattle washing facilities and canteen. It is a far cry from the few sheets of rusty galvanised iron that were in place when I first visited bull sales about 1965. At these sales venues in the fifties and sixties the inspectors from the Department of Agriculture were once more in evidence. They inspected all the bulls again and those that were above average quality were awarded premiums. What these premiums meant was, that if the purchaser of a bull was buying for "the country" rather than his own use, he was paid a premium of £50.00 per year approximately, because he was keeping a bull of higher quality. This often meant that a bull that left the Show Ring without a rosette in sight would be worth 100 guineas (£105). If he had a premium by the time he got to the sales ring he could be worth 150 to 200 guineas.

Keeping a bull for "the country" was not all plain sailing, sometimes it was difficult to get paid for "services rendered". The usual procedure was, that rather than pay for each cow at the time, an account was kept and the owner paid at the end of the season. I knew of one farmer who kept a bull for the country. The first year a neighbour brought twenty cows but at the end of the season failed to pay. The following season the same twenty cows were brought, same story no payment. The third season he started to come again but this time he started to bring heifers that were from the first years mating. At this point the bull owner called "stop the lights". In fact as a result of this he decided to no longer keep a bull for "the country".

Another farmer in the area normally kept a "Shorthorn Bull for the Country". On one occasion he decided to buy an Aberdeen Angus Bull. I mentioned in an earlier chapter that Aberdeen Angus were a small breed. This presented a problem, when his neighbours came with large cows, his bull was not able to reach them. The solution was to dig a hole and stand the cow in the hole with the bull standing on the higher ground. At the end of the first breeding season he was fed up digging holes. He decided to sell his Aberdeen Angus bull and buy a Hereford instead. At the start of the next breeding season he inserted an advertisement in a local paper "have sold small bull, bought big bull, can now bull cows without any hole". Hold on a minute my wife is constantly reminding me "to the pure, all things are pure".!!

The farmers using these Country Bulls also had their problems. Firstly, their choice of bull was limited to whatever was within walking distance. This could be a Shorthorn going one direction and a Hereford going in the other. As I said, the cow had to be brought on foot. You put a rope halter on her and away you went. If the cow had been there before she would probably run all the way there, but coming home you had to drag her, kicking and screaming. If it was a young heifer that had not been to the bull before, she had to be dragged kicking and screaming both ways!!

All this began to change in the mid to late fifties with the introduction of Artificial Insemination (A.I.). In the beginning there were only a few A.I. stations operating. However the farmer still had the choice of perhaps four or five breeds for the price of a phone call. Nowadays of course there are a lot more companies offering A.I. with the result that the farmer is spoilt for choice.

The introduction of A.I led the way to a lot of other developments. One of the first of these was frozen semen. This meant that a farmer who wished to use a particular bull that was standing in U.S.A., Canada or New Zealand could have frozen semen flown in.
We also got "heat synchronization". This was a big plus for the farmer who kept his cows for suckling (beef production), rather than milk. With sucklers heat detection is much more difficult. The development of "Hormonal Implants", meant that a batch of cows were ready for insemination on the same day.

Another major advance was "E.T.". No I don't mean the little green fellow with the big head. In this case E.T. stands for Embryo Transfer. With this development it was possible, if a farmer had a particularly good cow to get several calves per year from her. At the early stages this involved inseminating "Pedigree" or "Top Notch" cow and having a recipient cow also on heat on the same day. One week later the fertilized egg or embryo was removed from the Donor Cow and transplanted into the recipient cow. The recipient cow then carried the calf for the full gestation period. However, other than supplying the nutrients necessary for the embryo, the recipient cow had no influence whatsoever on the calf. Its entire genetic make up had been supplied by the donor and that cannot be changed. A few weeks after the transfer, the donor cow came into heat again and the whole procedure was repeated. In other words, the farmer was able to get perhaps four or five calves from the donor cow in a year. This allowed him to increase the genetic value of his herd. Alternatively, he had high genetic value calves to sell.

Now science has moved on even further. Firstly the donor cow is treated with hormones, which ensure that she comes on heat on a given day and also that she will release a number of eggs for fertilization. One week after the hormonal injection the cow is ready for insemination, and one week later her eggs or embryos are ready for harvesting. She may have only one or two eggs but she could have as many as seventeen or eighteen. Here again science has advanced. After harvesting, these eggs can be examined in the laboratory and can be identified as male or female. These sexed embryos can either be transplanted into a recipient cow or may be frozen and stored for future use.

Another big change in A.I. at present, compared to the 1950's, came about in the late seventies. This was the introduction of D.I.Y.-A.I. As a result of D.I.Y., conception rates were greatly improved. The problem with A.I. prior to this, was that the operator worked 9am-6pm while nature operated 24/7. If a farmer noticed a cow in heat when he was driving the cows in for milking at 5.30-6pm, he could not get the A.I. operator until next morning and by then the heat might have gone off. With the D.I.Y. system the farmer could inseminate the cow after milking. The new system also gave the farmer a better control over the bull he used. He could buy the straws he needed in advance and keep them in Dry Ice. Under the old system , by the time the operator got to your farm (if it were in afternoon), he might not have any straws left for the particular bull you required and so you had to settle for second choice.

I should have said earlier that when A.I. was first introduced, breeders of pedigree bulls were very worried. They felt it was the end of the Live Bull Trade as they knew it. In fact I knew of one breed society who took this threat so seriously that they refused to accept for registration any calf that had been conceived by A.I. Fortunately this fear did not last for very long and things soon settled down. Today there are more bulls in use than ever before in spite of an increased use of A.I., this is partly due to the increased number of cows in the country. However, it is also due to the fact that most large dairy farmers, having used A.I. on their better cows will, towards the end of the breeding season, run a bull, usually from one of the beef breeds, with their cows to "Mop Up" the stragglers.

Sheep - Sheep have not availed of science to the same extent as cattle. For the most part breeding is still by natural mating. The main reason for this is that heat detection with sheep is very difficult to spot. In fact even the RAM can miss it. However, science did not pass them by completely. Since the late 1960's heat synchronization or "Sponging" has been used in some of the larger flocks. This enables the flock owner to have a number of sheep lambing or yeaning at the same time. If left to their own devices, lambing could stretch from January through to May, with sponging this can be tightened up to about six to eight weeks.

Another advantage of "Sponging" is that you can use A.I. in conjunction with it. However it is still very much a specialist operation. In exceptional circumstances, E.T. is also practiced. I have heard of a Hampshire Down breeder who, while on a visit to New Zealand, was most impressed with the size and quality of the sheep out there. He thought of flying home some A.I. straws. However, as this would only give him the male side of the equation he decided to go for an E.T. as this would give him the female side as well.

Chapter 26

A look at our neighbours (near and far)

Our nearest neighbours are just up the road in Northern Ireland. Here the type of farming and the methods used are similar to our own. The farm size would be much the same as our own. However, there are a few differences. As I mentioned in an earlier chapter, they continued to grow flax long after we ceased to do so. They also have a lot more Ayrshire cows in their dairy herds. This is evident by the fact that at Balmoral Agricultural Show in Belfast, every year they have a section for Ayrshires.

For the purpose of disease control, movement of livestock between Northern Ireland and the Republic is very much restricted. Cattle, sheep or pigs can only move between the two countries if they are going for slaughter. If livestock come from Northern Ireland to attend a Show down here, they cannot return to Northern Ireland. The same applies to stock going to Northern Ireland from here. The exception to this ruling is all equines (horses, ponies or donkeys). They can move backwards and forwards quite freely. In fact every year, at Bailieborough Agricultural Show, up to 50% of our equine entries would be from Northern Ireland.

If we go across the water to Great Britain, we have similarities and exact opposites. In Scotland, types of farming and the size of farms would be much the same as here. Quite a number of Scottish farmers travel to Ireland every year to buy Store Cattle. In the past the connections between Scotland and Ireland would have been strongest back in the 40's and 50's and even into the 60's when there were a lot of potatoes grown in Scotland. Hundreds of small farmers (small in acreage, not stature!) from Donegal took the boat to Scotland in May/June for the potato harvesting. This was known as "Tatty Hoking" and by moving from one area to another they could get work right through to October.

The connections with Wales would be similar in many ways. The size of farm and type of operations are much the same as in Ireland. Most of the live cattle exports to the English Midlands would have been shipped through Holyhead in Wales. In England the situation would be much different. While the type and methods of farming would be similar to Ireland, the size of enterprise would be much larger.

However, the main difference between farming in Ireland and farming in Britain in general, but England in particular, would be the ownership of the land. In Ireland practically all land is owner/occupied having bought out the landlord. The exception to this would be temporary leasing. For years this would have meant the "Eleven Month" system, where land was rented, for example, from 1st January to 30th November. In more recent times we have seen 3Year Lease or 5 Year Lease. Across the water in England, while some land is owner occupied, a large proportion is owned by Banks, Insurance Companies etc. It is quite common to find the third and fourth generation of the one family renting and living on the same farm.

Now let us go a bit further afield to North America. One normally thinks of America as being bigger and better than everywhere else. America also has the reputation of being at the forefront of scientific and technological advancements. Well, not always the case!!!!

Back in 1995 one of my daughters was living and working in Portland, Maine. While on a visit to her we attended a "County Fair" (we would call it an Agricultural Show) at a place called Farmington near the Canadian Border. Over two days I spent a lot of time with a local family who were showing prize winning Holstein/Friesians. I learnt that they had a herd of 40 high yielding cows. However, I was amazed to learn that these cows were housed in a "Tie up Byre". They were milked in this byre with a bucket unit (see chapter on Dairy Cows).

The silage to feed these cows was brought in with a wheelbarrow, and the waste at the other end was brought out in a barrow (not necessarily the same one). Speaking of silage, they told me that they had made "Big Bale Silage" for the first time that year.

Now to the point of my story. In Ireland in the mid nineties, any farmer with 40 good quality cows would have them housed in a loose cubicle house and they would have been on "self feed" or "easy feed" silage. Also we had big bale silage here since the early eighties. See chapter "Nothing New".

Going back to that "County Fair". I saw two things of interest in the sheep pavilion. Firstly, I saw the largest sheep I had ever seen. They were the size of donkeys. The breed was "Colombian". The second thing was, sheep with canvas coats on them. I could not understand the reason for this, as the weather was quite warm at the time. On enquiring, I was told that the ones with the coats were wool breeds rather than meat breeds. When they went out to the show ring the quality and appearance of the fleece was all-important. The coat had two functions, firstly it prevented the spectators from poking the sheep and thus messing the fleece, secondly, the coat kept the sheep very warm and when the coat was removed just before going out to the show ring, the wool was more pliable and easier to get it fluffed up properly.

Now for the other side of the American coin. In 2002 a friend of mine sold his house and land on the Cavan/Meath border and went to America. He bought less than thirty acres of land. In a short time he had houses up to accommodate and milk 200 cows. Today, four years later, on that same piece of land he is milking 800 cows. Needless to say the cows never get out, and all feed, winter and summer, has to be bought in. In fact my friend rarely gets to actually milk any of his cows. He is kept busy buying and organising delivery of feed. Towards the end of 2005 he wanted to put in a new 50-cow rotary milking parlour. He got a better deal from a company in County Kerry to ship over the equipment and also the men to install and set up, than he could from any local firm.

Our next port of call is to the other end of the world, New Zealand. The land in New Zealand is very similar to Ireland; some of it is very good and some very bad. The weather is much the same. However, there the similarities end. Some of their farms run to thousands of acres. In fact they would not consider anything less than 150/200 acres as being viable. Having said that, I should point out that there is a new development taking place at present. This is where some of the larger farms are selling off part of their land in small sections of about 10 acres. These are being bought by families from the nearby cities who want to get away from the smog and noise of the city "Rat Race". On these "Life Style Blocks" as well as keeping a pony for their children, they also keep some hens, a few sheep and maybe even a pig or two. I should have said that the "Bread Winner" continues to commute to the city every day.

Now back to the mainstream farming. While there are some dairy herds with as few as 50/60 cows, 1,000/1,200 cow herds are quite common. Sheep flocks could run from 200 to 5,000 ewes.

On my first visit to New Zealand in 2002 I visited a dairy farm with 1,200 cows milking through a 50 unit rotary parlour. This parlour was the only farm building about the place. The cows were outdoors over winter and calved outdoors in small paddocks. Any calves not required for herd replacement were sold at about 3-4 days of age. The ones being kept were moved to another farm where they were reared. They did not return to the home farm until they were in calf.

I came across two other things of interest concerning the dairy industry on that trip. Firstly, I saw a number of Friesian/Holstein herds with a Jersey Bull running with them. Nobody could explain to me the reason for this. The second interesting thing was, where two farmers used the one milking parlour. This parlour was built on the boundary of the two farms. At milking time, farmer "A" milks at 5 am and 3 pm while farmer "B" milks at 7.30 am and 5.30 pm.

The advantage of this arrangement was that they each had less capital tied up in buildings.

I also visited a sheep farm carrying 3,500 ewes. Unfortunately they were in the middle of shearing on the day of my visit and so nobody had time to talk to me. The ironic thing was there were more buildings on this sheep farm than on the dairy unit visited. All sheep farms regardless of size will have a Shearing Loft, a number of Holding and Sorting pens, plus a shed to store the wool.

Another interesting fact I learned was concerning the Possum – a possum is the shape of a rat but the size of a rabbit. This possum causes havoc in two ways: the amount of grass they eat and the spread of bovine T.B. I cannot remember the figure for the estimated number of possum in the country but I read an article stating that if all the possum came together in the same area, in one night they would eat 31,000 acres of grass!!! Some years ago it was proven fairly conclusively that the possum was responsible for spreading Bovine T B. Now in Ireland, when it was proven that the badger was spreading Bovine T.B we were told we could not touch the poor badger. Not so in New Zealand. When the connection was first identified some years ago, they came out with the Army and virtually shot anything that moved. Nowadays if there is a T.B. breakout they put down poison all over the surrounding area.

My next visit to New Zealand was in 2005. Shortly after arriving I visited a Ram Sale. This was very interesting as there were a number of breeds there that I had never heard of. There were a couple of self-moulting breeds, in other words they never have to be shorn. However the breed with the largest entry was the Suffolk.

All the rams for sale were Shearling or Hogget rams (one year old). There was a minimum fixed price of 700 NZ Dollars (€470 approx). They ranged in price from 700 dollars to 4,000 dollars. The Suffolk also had some Hogget ewes entered. They ranged in price from 250 dollars (€165) to 500 dollars.

The one thing all the rams, regardless of breed, had in common was small fine heads. In Ireland we would say they had ewe's heads. When I remarked on this to a farmer, who was there to buy a ram, he told me that they had spent the last ten years breeding them that way for easy care lambing! In other words there would be fewer problems at lambing time if the ram used had a small head. This is very important if you have a few thousand ewes and very little help.

On this question of labour availability, I learned that some sheep farmers have been forced out of business, because they could not get workers, or if they could get them they could not afford to pay them. The alternative is to sell their sheep and plant the land with trees instead.

On the dairy side, the love affair with the Jersey bull seems to be stronger than ever. In all the dairy herds I saw most of the cows were Jersey or Jersey cross. The Jersey breed is famous for producing milk with very high butterfat content. For a number of years now health authorities worldwide have been condemning the production of butter fat.

I also discovered that New Zealand had a quota system for milk production. This quota system is instigated by Fonterra, the largest milk processor. In order to have your milk collected by Fonterra you must buy shares in Fonterra to cover your production. If you have a particular good grass-growing year, and therefore produce more milk, you must either buy more shares to cover this extra production, or pour it down the drain. These shares do not come cheaply. I read in a farming paper out there about one farmer who had been supplying a small independent creamery. That small company had now been taken over by Fonterra and so he was obliged to buy Fonterra shares in order to have his milk collected. The newspaper did not say how many cows he was milking but they figured it would cost him 400,000 NZ dollars to buy his "Quota".

Turning to beef production. I learned that the vast majority of beef cows were either Aberdeen Angus mated to a Hereford Bull or Hereford cow mated to an Aberdeen Angus Bull. I did see a few Charollais but they would be few and far between.

While touring on the South Island I came across more deer herds than beef herds. The system of production is different than in Ireland. Here the deer are normally born, reared and finished on the same farm. In New Zealand it is common for one farmer to breed them and rear them, but then sell them as stores to another farmer to finish.

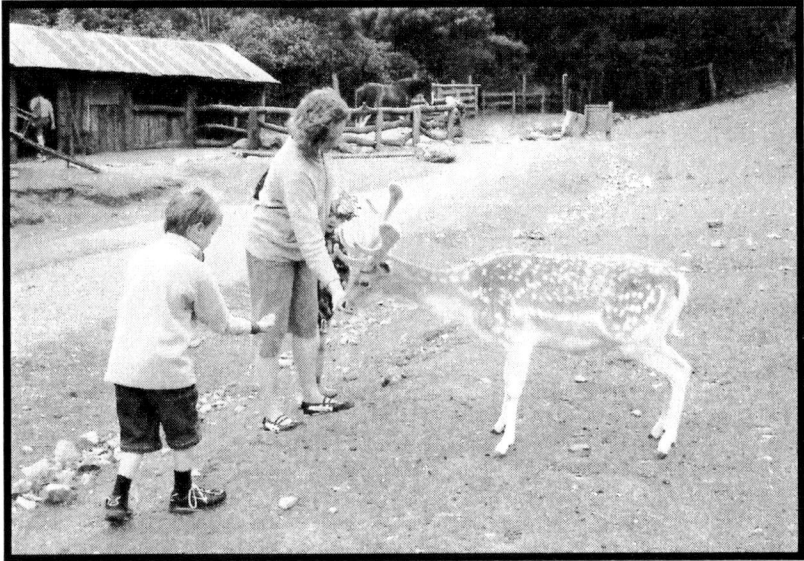

"As can be seen, deer in New Zealand are tamer than in Ireland!!! Photo taken in Staglands Park, Upper Hutt, showing my daughter and two of my grand children"

Another thing I came across on the South Island was the number of farmers who were using irrigation systems. November 2005 was a particularly dry month. By Christmas anyone who did not have an irrigation system was in serious trouble. I read in some of their farming papers that the water levels in New Zealand in general are dropping to such a degree that it is causing grave concern. They reckon that in about five years time, if things continue as they are at present, that they will be in danger of drawing up salt water in their bore holes. The proposed solution is to set up dykes, as in Holland, to slow down the flow of the rivers into the sea and therefore keep more water in the country. The only problem is, that the estimated cost of all this is 28 billion dollars. We think we have problems here!!!!!!!

Chapter 27

Things we don't do now

There are a number of things we did in the past that we don't do now. These include milking cows by hand, except in an emergency, having to light an oil lamp to go out to look at our animals after dark, going to the well every time we need to get a bucket of water. However there are other things that might not come to mind so readily.

Animal Dosing

Nowadays, most anthelmintics are liquid and ready to use. Back in the forties, fifties and even the sixties if we wished to dose a bullock for worms, we had to reach for the Phenothiazine powder. We then had to dissolve the correct amount (depending on size of animal) in water, fill the mixture into a bottle, and then administer it to the animal. If you had ten cattle to dose, you filled the ten bottles before you started. If we wished to dose for fluke it was the same story. However this time we used Hexachloroethane. When we had to dose for both worms and fluke, we bought a mixture of Phenothiazine and Hexachloroethane.

If we wished to worm either pigs or poultry, in the past, we had to use Piperazine Citrate. Once again this came in powder form. However, unlike the case with cattle, with pigs and poultry we just added it to the drinking water. Nowadays things are much simpler. The first development in the late sixties was the "ready to use" liquid drenches. This would be the most common method used today. Then in the late seventies, we got the injectable anthelmintics. The most up to date is the "pour on". With this method you just pour the treatment along the centre of the animals back. This firstly kills the external parasites, like fleas, ticks etc. Then it is absorbed into the blood stream from where it kills the internal parasites.
Isn't modern science wonderful!!!!

Drainage

Because Ireland gets so much rainfall, the land is always very wet. For this reason, Irish farmers are constantly trying to drain the land. In the "Dark Ages", before my time, this was all done by hand. Drains or shores were dug across the field with spades and then these drains were filled with stones. The surplus water ran through these shores until it came to the main drain or "shuck" at the edge of the field. Provided these "Shucks" were kept clean the water then ran from there to the nearest river.

Sometime in the late forties or early fifties, the earthen-ware pipe was invented. These were about 12 to 15 inches long. They were made in Kingscourt and were delivered to the farmer by the lorry load. The shores still were, for the most part, dug by spade. The pipes were brought to the fields by horse and cart and dropped off at intervals along the length of the shore. Then the "hard labour" began. These pipes had to be dropped into the shore, one at a time. If you dropped them too close together and they hit off each other they would break. You then had to get down and remove the broken pieces and then drop another one. This job was often done during winter. On a frosty morning the pipes would be stuck together. This left it very hard on one's hands. At 12 inches per pipe you can imagine the number needed to fill several shores across a large field. Nowadays you just get a 100 metre roll of Plastic Land Drainage pipe and roll it out into the shore, which has been opened by machine.

Ghost Drovers

I mentioned in an earlier chapter about cattle bought at the Fair in Bailieborough, going to the Dublin Cattle Market. Most of the cattle there were bought for export to England. To get from the Dublin Market (Prussia Street) to the Docks at the North Wall, the cattle were normally walked through the streets of Dublin. They would be driven in batches of 20 – 25 cattle at a time. Herein lies my story. Men from the back streets around Prussia Street would present themselves for the task of driving these cattle. This was OK back when there was no motorised traffic, but as cars, buses etc began to get more plentiful in the late fifties and early sixties, these "Dubs" found themselves unable to handle this job anymore. However, as it had always been their job by tradition, they would not allow anyone else to do it. Eventually the cattle exporters came up with an idea. They paid these men to stay at home. They then brought in good cattle men from the country. Two good cattle drovers and one dog could do a better job than five or six of the locals hence "Ghost Drovers".

Stone Walls

There is a district between Bailieborough and Virginia known as Stonewall. As the name would suggest, the fields were divided by stonewalls rather than hedges. These walls were originally built with the stones that were gathered from the fields. When they were newly erected, they were not only colourful looking but were also quite effective as a means of containing livestock in one field at a time. However when these walls began to crumble and fall down they were very difficult to repair. The original break in the wall might be quite small, but in order to repair it one would have to dismantle quite a large area before one could start to rebuild. Suffice to say that it was a job that was hard on the fingers and on the back. Thankfully, a lot of these walls have now been replaced with wire fences.

Waterproof Clothing

Nowadays, if we have to work outdoors, in inclement weather, we put on our " oilskins " or rubber/plastic pull-ups etc. In the 1950's we had to make our own. To do this we got a white flour bag (it would have held 50kgs flour). We opened this out and stitched on straps, at one end to go across our head, and in the middle to tie around our waist. We then painted the bag with linseed oil. This was repeated over several days, until the bag was so stiff that it would stand up on its own. It was now completely waterproofed. When worn, this "apron" came down past the top our wellington boots, and so, our fronts at least were fully protected.

Chapter 28

Miscellany

Farm Advisory Services

My earliest memory of the Advisory Service was in the Mid Fifties. Pat Mc Hugh was the advisor in the Bailieborough area. It was he who persuaded me to attend Gurteen Agricultural College. At that time the advisory service was under the control of the County Committee of Agriculture. This illustrious body was a group of the most up to date farmers in the County. They were appointed by the County Council. The County Council paid the salaries of the advisors.

At that time every County had its Committee of Agriculture, and accordingly its team of Agricultural Advisors (B.A.g.'s). These B.A.g.'s were all trained in the same place, namely the Albert College in Glasnevin. At the time the principal of the college was a very colourful gentleman named Professor JD Ruane.

The County Committees also employed Poultry Instructors (P.I.). These were trained in the Munster Institute in Cork. As I said, while these B.A.g.'s and P.I.'s were trained under the same set of standards, when they got out on the ground, they did not all sing from the same hymn sheet. The various County Committee's of Agriculture were advocating the best possible farming practice, at that time .To the best of my knowledge there was no co-ordinating body between the various Counties.

Towards the end of the sixties, An Foras Taluntais (The Agricultural Institute) was set up to undertake Agricultural Research. Research stations in the various disciplines, Dairy, Beef, Sheep, Tillage etc. were established at strategic positions around the country. This was all under the control of the Department of Agriculture. This research continued alongside the County Committees until 1980. At that time a new organisation A.C.O.T was formed. This stood for An Chomhairle Oiliuna Talmhaiochta, meaning The Council for the Development of Agriculture.

This took over the role of the County Committees and for the first time we had a co-ordinated advisory service. This meant it did not matter if an advisor was placed in Dublin or Donegal, Cavan or Cork. They were all advocating the same policies. We then started to get Specialist Advisors. Instead of all advisors being all things to all people, we now had the establishment of Dairy Advisors, Pig Advisors, Accounts and Taxation Advisors.

Agricultural Education also came under A.C.O.T. Those colleges that were owned by the Department of Agriculture were signed over to A.C.O.T. Private Colleges, while they still controlled their finances etc., now had to work to the syllabus laid down by A.C.O.T.
A.C.O.T. continued to work alongside An Foras Taluntais, which was still operated as a separate entity. Then in 1988 the present organisation Teagasc (Agriculture and Food Development Authority) was set up. Finally, Research, Farm Advisory and Education are under one umbrella.

In the sixties, some of the co-operatives started employing Agricultural Advisors. Depending on the activities of the Co-Op, these would be specialists in Dairy, Pig or Tillage.

Bailieborough Agricultural Show

In 1965 Bailieborough Agricultural Show was started. At first it was confined to horses, ponies and donkeys. Over the years it has been built up, until today (2006) where it includes Horses, Sheep, Dogs, Cattle, Flowers, Vegetables, Cookery, Art and Photography. It is now recognised as one of the best shows in the North East.

Protein Feed Blocks

In the mid sixties we got protein feed blocks. These were a mixture of Mollasses and Urea with vitamins, minerals and trace elements. They contained about 20% protein. The idea behind these blocks was that you put them out in the field, in winter, with either cattle or sheep. When the animals licked at these blocks, the energy provided, encouraged them to forage over a wider area, looking for food. It kept the livestock more contented. The blocks were recommended for use with sheep out on the hills. I used them for a few years myself with Suckler Cows and found them quite effective. If I was late getting out with their hay in the morning, the cows, instead of standing at the gate roaring, were scattered about grazing.

However, the Agriculture Institute at Grange in County Meath carried out tests on these feed blocks and said they were not cost effective. As a result they were removed from the market within a few years.

N.P.A. (National Ploughing Association)

The N.P.A. was set up around 1930. In the beginning it was a small affair, with only horses ploughing. However, by the late forties tractor classes had been added. In the late fifties ladies classes were also included.

Gradually over the years this event has continued to expand and is now a National Institution. At the present time (2006), it is the largest Annual Agricultural Event in the whole of Ireland, North and South.

Funny thing to tell, I last visited the Ploughing Championships in 1962-63. I had all plans made to attend again in 1966, but the day before it started, I finished up in Baggot Street Hospital in Dublin, with a suspected stomach ulcer. However, maybe it was as well that I did not attend that year. I read in the newspapers, and heard on the radio during that week, that the weather was extremely stormy. The canvas tents of the officials and the trade stands were blown away, some into the next Parish and some into the Parish after that again.

Down through the intervening years, I have made several attempts to attend, but for various reasons never made it. Either I was too busy at work, had just come back from holidays and therefore could not get time off, or for some other reason. However, I plan to retire in 2007 and the N.P.A. will be high on my list of things to do.

To continue with the N.P.A. the County Cavan Branch was formed in 1980. Sam Clarke from Killinkere was one of the founder members, and is still going strong. The branch holds county championships every year. They also have to supply Stewards for the National Event.

I.C.M.S.A. (Irish Creamery Milk Suppliers Association)

The I.C.M.S.A. was founded in 1949. One of its most stalwart supporters in the early days was the late Charles Fletcher, who later became Chairman of Killeshandra Co-Op. Today the ICMSA is the second largest Farmers Organisation after the IFA.

I.F.A. (Irish Farmers Association)

This was originally called the N.F.A. (National Farmers Association), but the name was changed in 1955. As we were starting to look to Europe it was considered more appropriate to have the name "Irish" included.

Silage Making

In the later 1950's farmers started to make Silage. At first this was made in Clamps. However there was too much waste with this method and so Silage Pits, with sloping walls and sloping floors were erected. Preservation was much improved and therefore there was less waste.

This suited the larger farmer but it was not suitable for the farmer on a smaller acreage. In the early 1970's "Baled Silage" was tried for the first time. The first attempt was to bale the fresh grass in the usual way using a "Square Baler". These bales were then arranged in a clamp and covered with polythene and sealed in the normal way. This method was dead before it was born because the Agricultural Contractors thought that it would damage their machinery. Some time later the "Round Baler" was invented. This was much more suitable for "Baled Silage". However, we still had a problem; these bales had to be wrapped. The initial method was to drop the bale into a large plastic bag. This was very laborious. Then in the early 1980's we got the Bale Wrapper. This works on the same principal as shrink wrapping any other product at the point of manufacture. At present there would probably be as much silage made by this method as the traditional "Pit Silage".

Fertilizer

My mind goes back to 1955, when as a 14-year-old boy I struggled to carry in bags of North African Phosphate from the delivery lorry into one of the sheds in Graham's Yard. This material came in 16 Stone = 100kg Jute Sacks. It was known as Super Phosphate. We also had Muriate of Potash. This was in more reasonable 50kg Paper Sacks. Another product at the time was Basic Slag. It was a very fine black dusty material. When spreading this it got into your hair, your ears, up your nose, down your throat, into your navel or any other crevice it could find!! I cannot recall having any nitrogen in those far off days. I do remember that we did not call them fertilizer; no, we called them artificial manures.

Tractors

The earliest tractors in the 1930's – 1940's were very simple compared to what are available today. Firstly they were driven by petrol, the diesel engine had not yet been invented. Even at that time petrol was very expensive. The next development was T.V.O. (Tractor Vaporising Oil), with this system you started your tractor in the morning, when cold, on petrol and then when the engine warmed up you switched it over to the main tank which contained T.V.O. These early models did not have P.T.O. (power take off) or Hydraulic Lift Arms. In other words all they did was pull the machine being used in the same way, as it would have been pulled by a pair of horses. You were dependent on the Ground Wheel Movement to drive the mowing machine or whatever was being used. In some cases one might as well have been working with horses, because it took two men, one to drive the tractor and one to work the machine behind.

Ferguason 20. (over 50 years old). At controls - two of my grandsons

In the early 1950's a man from Northern Ireland came to the rescue. This was the now famous Harry Ferguson. He had developed the concept of the P.T.O. and Hydraulic Arms. Having tried to sell these ideas both to Ford Motor Company and David Browne, without success, he then sought finance to start his own company. With great difficulty he succeeded in doing this. He developed the Grey Ferguson 20. It had a diesel engine, P.T.O. and Hydraulic Arms and was just the ideal size for the small County Cavan Farms. There are some of these tractors still in full use, even though they are over fifty years old. Shortly after this, Ferguson was taken over by an English company called Massey Harris. Afterwards it traded as Massey Ferguson, and as they say "The rest is history".

Maize

In the early 1990's a new crop was introduced to Ireland. This was Maize. Prior to this the growing of maize was associated with Countries that enjoyed a warmer climate than Ireland. In fact in the first couple of years in Ireland, maize was a disaster. Then someone came up with the brilliant idea of growing it under polythene. This was a big improvement as it helped to keep the heat in the soil. It is now quite common to see, large fields with polythene stretched across them. Well in case you did not know it, this is a crop of maize. Lest any conservationists among my readers might be worried about this, let me assure them that the polythene being used is bio-degradable. When the maize has been harvested, what remains of the polythene is ploughed in and it continues to disintegrate underground.

Quads

A modern piece of machinery, which has made a vast improvement to working conditions on some farms is the Quad or A.T.V. (All terrain vehicle). Whether it is for rounding up the cows for milking, bringing a few bales of hay to outlying cattle, driving to check sheep on a hill field or bringing a ewe and her new lamb in from that same hill field, it is invaluable. Because of its lightweight it was possible to use it to spread fertilizer when the ground was too wet to carry a tractor.

Donkeys

I have mentioned more than once during this book about various Grants or Premiums having been paid to farmers. The one animal that never received a grant was the humble donkey. This situation was nearly rectified in the 1960's. The Minister for Agriculture at the time, Mr CJ Haughey, promised to pay out a grant to any farmer who could train his donkey to go "Haw He" instead of the usual "He Haw"!! I don't think it was ever paid out!!

Ringworm

This is a viral infection of cattle and to a lesser degree horses. It was widespread in the 1940's, 1950's and into the 1960's. The parts of the animal most affected, were normally the head or neck. At first a few small scabs would appear, but if left unattended it spread until the entire head and neck was covered. These scabs were very itchy. The poor calf then scratched its head on the feeding trough, hay rack or even the frame of the door. Therein, lay the problem, when another calf touched the trough, hay rack etc, it picked up spores and the disease was spread again. In my youth there was very little on the market to treat Ringworm. The normal way of trying to control it was to paint the affected area with Waste Oil. Sometimes this worked and sometimes it did not. Then in the 1960's a product came on the market, which was very effective. This had to be added to the calves feed and fed everyday for about 2 weeks. The spores of the ringworm are very difficult to kill. In fact the only sure way to clear an infection from the calf house was to "burn" all surfaces to which calves had contact. The simplest way was with a blow-lamp, (normally used for stripping paint). While this was effective it was always dangerous as well, due to the fact that calf pens, hay racks etc., were all made with wood.

There was another very serious aspect to Ringworm; this was that it could be transmitted to humans. The problem then was that there was very little the medical profession could do for Ringworm. In the end the sufferer had to go for "The Cure".

Nowadays Calf Pens are normally constructed with tubular steel. The same thing applies with Hay Racks etc. This makes for easier cleaning and disinfecting of pens between batches of calves. Add this to improved Livestock Husbandry in general and we now have a situation where one rarely sees Ringworm.

Thatching

Another craft that is very much in decline is thatching. In my youth there were a lot more houses around with thatched roofs. Nowadays, one rarely sees a thatched roof. This is a pity really. Apart from the appearance of a newly thatched house, it is a fact that the thatched roof gives better insulation and heat retention than any other type of roof. There are reasons for the decline in thatching. Firstly, the art of thatching has not been passed from father to son. Secondly, with modern methods of Tillage Farming, the type of straw produced is not suitable for thatching.

"Thatched Cottage"

Rape Seed

This is another crop that has gone through the changes over the years. Back in the 1940's and 50's rape was grown to be strip grazed by sheep. In the 1960's when harvested, the seed was used in animal feed. For many years it went completely out of fashion. Now it has got a new lease of life. It is being grown as a source of bio-energy. The acreage of rape now being sown is on the increase every year, and with the way the price of oil keeps rising, this is set to continue. For those not familiar with the crop, it is very easily recognisable in May/June by its bright yellow flower.

Chapter 29

Farmer Enterprise

In the 1960's the farmers decided it was time to take control of their own destiny. For the most part the only commodity controlled by farmers at that time was milk. The exception to this was the old Mullagh Co-op, which was set up as a trading co-op so that their farmer members could get their farm inputs at the right price. They also set up a Feed Compounding Mill. Later Mullagh Co-op became N E F (North Eastern Farmers). This would have been in the late seventies.

Now back to the sixties. Most Co-ops set up stores, then feed mills. Falling into these two categories locally we had Killeshandra Co-op, Lough Egish Co-op, Oldcastle Co-op and Kilnaleck Co-op and of course Bailieborough Co-op.

The next move was into the area of Livestock Sales. The first group of Co-op Marts was set up by Cork Marts. They set up branch marts all over the South. Towards the end of the sixties there was a nationwide Share Capital drive by Cork Marts to raise enough cash to purchase I M P (International Meat Packers) in County Kildare. This was the first venture by farmers into meat processing.

Also in the Livestock Mart business we had Golden Vale. They opened marts all over counties Limerick, Tipperary, and Offaly. They even came as far north as Carrigallen in County Leitrim.

In 1967 Ballyjamesduff Mart was opened. Shortly afterwards they opened their General Store. At present they are one of the largest assemblers of wool in this region.

Later in the sixties Clones Co-op Mart came on the scene. In the early nineties Cavan Co-op Mart was formed to take over Cavan Mart from Brady Brothers Coothill.

The mart in Granard is co-operatively owned, as was (I think) the Mart in Drogheda, which sadly is no longer with us.

I have mentioned all the co-operative marts in the locality. There was one other that nearly was – Bailieborough! Towards the end of the sixties, the late Tom Murtagh, father of Peter, was very keen that we should have a mart in Bailieborough. Meetings were held and an Ad-hoc Committee was set up to examine the possibilities. I myself was a commercial traveller at the time and on my journeys around the country I visited Marts all over the place. On visiting there I took measurements of Stock Pens, Sales Rings etc.

Eventually the committee felt confident enough to go ahead. A suitable site was purchased subject to planning permission, and licences being obtained. The former was obtained without much difficulty. However, the licence was a problem, this had to be obtained from the Department of Agriculture. The response from the Department was that with Marts in Kingscourt, Carrickmacross, Cavan, and Cootehill, the needs of Bailieborough were adequately catered for.

Negotiations were still ongoing, delegates visited Dail Eireann, letters were written and local T.D.'s were canvassed. However, while all of this was going on, Kingscourt Mart applied for and received planning permission and licences to build a Mart in Bailieborough, for pigs only. They proceeded to erect this on the Shercock Road, beside St Anne's G.A.A. Park.

Needless to say this took the wind out of the Bailieborough committee. Even when the Kingscourt Auxiliary Yard closed down, after trading for a few years, the Bailieborough Committee had disbanded and to the best of my knowledge never tried to reform.

Farm Relief Services

In the early seventies young farmers, especially dairy farmers got tired of being tied to their farms 24/7. As a result Relief Milking Clubs were set up. Under these clubs I milked for you this week-end to allow you to go to the football match and then you milked for me next week to allow me go to a wedding.

Towards the end of the seventies Macra Na Feirme (Young Farmers Club) saw the merits of these clubs and realised that the concept could be acted upon. Out of this, Farm Relief Services came into being. Again this was set up by farmers for farmers. I mentioned in an earlier chapter how this operated.

Co-operative Pig Fattening Units

When writing about pigs I mentioned about Co-operative Pig Fattening Units being set up. For the most part these were started in the early to mid sixties. They were built in parts of the country where there was no tradition of pig fattening. These units were a great help to the local farmers who had sows. Instead of having to bring their bonhams to County Cavan or Monaghan to sell them they now brought them to the Co-op.

Cootehill Poultry Products

In 1956 this co-op was set up. They started with a hatchery supplying day old chicks to the farmers of the area. Later they set up a poultry processing plant, where they took in these birds, processed them, froze them and then distributed them around the shops. In the late sixties a similar co-op was set up in Monaghan. (1)

N.E.C.B S.

In 1951 Killeshandra Co-Op set up an A.I. Sub Station in Killeshandra. At first this had only two bulls, a shorthorn and a Hereford. Over the next 10 years, the use of A.I. increased and the range of semen also increased to include, Friesian, Jersey and Aberdeen Angus. Then in the early sixties N.E.C.B.S. (North Eastern Cattle Breeders Society) was set up. This was financed mainly by Killeshandra Co-op, Lough Egish Co-op and Town of Monaghan Co-op.

Their headquarters were initially in Ballyhaise but their Bull Stud was stationed in Cootehill. In the Spring of 2002 N.E.C.B.S was taken over by Progressive Genetics. In the Autumn of that year Progressive Genetics also took over N.W.C.B.S. Sligo. Some years prior to this Progressive Genetics was set up as a farmers Co-operative to take over the AI Centre run by D.D.M.B. (Dublin District Milk Board). This new co-operative is now one of the largest A.I. Companies, and have control of about 1/3 of the country. I mentioned in an earlier chapter (Livestock Breeding) that at the beginning of the sixties the various Cattle Breed Societies were worried that the days of the "live bull" were numbered. Fast forward to March 2006, and we have the Department of Agriculture worried that farmers are not making sufficient use of A.I. The Minister for Agriculture has promised moral and financial support to the A.I. Societies in order to reverse this trend.(2)

Leather Tannery

In 1950, Lough Egish Co-op got involved in a Leather Tannery being started in Ballybay. Prior to this all cow hides had to be transported to Portlaw, in Co Waterford.(3)

An Bord Bainne

Was set up in 1961. This took responsibility for export sales of all Dairy Products. In an effort to instil uniformity all butter was exported under Kerrygold Brand. (4)

Mc Cormack Skim Plant

In 1962 a skim powder drying plant was set up in Killeshandra. This was a joint venture between Killeshandra Co-Op and and Lawes Brothers, an English Company. This new company traded as Mc Cormack Products. Prior to this all skim milk was either brought home by the Farmers to feed to pigs and calves or thrown down the drain. With the arrival of compounded animal feeds, farmers were less interested in feeding skim to pigs. Also around this time the number of sows being kept was on the decrease. By having the skim plant Killeshandra Co-op were able to pay a better price for "Whole Milk".(5)

Glaxo

In 1966 a similar joint venture for skim drying was set up between Lough Egish Co-op and another British Company – Glaxo.(6)

Springcool U.H.T.

In 1966 Killeshandra Co-op, developed its U.H.T. plant. Under the system little "Jigger", packs of milk and cream were produced which would keep "fresh" for a long period of time without the need of refrigeration. This was the first plant of its kind set up in Ireland. Nowadays "Jiggers" from Killeshandra are used for Airlines, and Hotels, all over the world.(7)

Butter Mini Packs

In the late sixties Lough Egish Co-op were trying to come up with a Value Added Product. To the best of my knowledge they were the first company in Ireland, to come up with butter mini packs. This was in 1968. Once again these mini packs are used by airlines and hotels all over the world. I mentioned in an earlier chapter how Grahams Hotel in Bailieborough used to churn their own butter and then bring to the table in little balls or pats. Lough Egish mini packs are a lot simpler.(8)
* Item No's 1-8 – Mac Donald 96

Bailieborough Co-op

While all this was going on, Bailieborough Co-Op was also on the move. Let us step back a little further in time; Bailieborough Co-op was first founded in 1902. When the idea was first put forward to form a co-operative the local farmers were receptive to the idea. Bailieborough had been forward looking for some years at the time, with regard to things agricultural. The reason for this was that over thirty years previously the first Agricultural Training School in the British Isles had been set up in Bailieborough. This was a residential school and was built on the site of the present Mental Health Clinic (previously Teagasc Office).(15)

While doing some research for this book, I came across the following statistics. In 1901 the total milking cow population in County Cavan was 47,772, by 1926 (Post World War 1), this had dropped to 42,837, however, by 1960 this had recovered to 45,000 and by 1974 had increased again to 55,000.Sheep numbers were as follows: 1901 – 27,103, 1926 – Dropped to 22,952, 1960 – 38,800 but by 1974 had dropped again to 36,200. Finally Pigs: 1901 – 52,746, 1926 – 39,846, 1960 – 57,200, and 1974 – 73,400.(16)

Back to the early days of Bailieborough Co-Op. In the first year milk price was less than 4 Old Pence per Gallon, Butter was sold at 10 Old Pence per Pound and daily intake of milk was approximately 550 Gallons.(17)

In 1913 the Creamery Manager sent a very passionate letter to milk suppliers enticing them to join the Cow Testing Association. (This was a forerunner for the present day Milk Recording Scheme). It is ironic that Dairy Farmers today are deliberating over whether or not they can afford to have their cows recorded. The arguments put forward by the Manager in 1913 are as relevant today as they were then.(18)

By 1915 the average amount of milk supplied per farmer (not per cow) had increased to approximately 900 Gallons. A record price of 5.07 Old Pence per gallon was paid.

In 1922 Mr Michael Fay was appointed Creamery Manager, and continued in this position until his retirement in 1964. Incidentally Mr Fay's son Paddy qualified as a chemist and ran a successful Pharmacy shop in Trim, Co Meath. (19)

By 1950, Milk intake had reached 500,000 Gallons, and by 1960 it hit 1,000,000 Gallons.(20)

In 1964 Paddy O'Brien was appointed Manager. Also in that year the milk intake was such that they could not cope with it. In order to relieve this situation a new separating station was erected at Mullagh. Its first Manager was Mr John Cooney. (21)

The year 1965 was an eventful one. Firstly Milk intake reached 2,000,000 Gallons. It was also the year that an evaporation plant was set up in partnership with Mc Cormack Products, Killeshandra. This was started up to part dry the skim, and so reduce the cost of transporting to Killeshandra. Every 1,000 Gallons of skim that passed through the plant, could be reduced to 250-300 Gallons by extracting the water.(22)

During 1965 we saw the retail stores set up in Henry Street, Bailieborough. This is now known as J & L Stores. In 1970 a store was opened in Townley Hall, Drogheda. This was followed over the next few years with stores in Arva, Ballyhaise and Navan.(23)

In 1970, bulk collection started from the larger suppliers. By 1977, 95% of collection was by bulk tank collection. (24)

In the early 1970's Bailieborough Co-Op Engineering Society was set up. This company was set up to manufacture and install stainless steel tanks and vats for Farm use, also in creameries and other food outlets. One of the largest orders, ever received for Vats was from Guinness Group when setting up the new Harp Lager Factory in Dundalk. By 1974 business was so buoyant that a second factory had to be built. Within a few years 50% of the output was being exported to the E.E.C. and Middle Eastern countries.

During 1976 another new company was formed to handle import/export trading with the Middle East and Third World developing countries. In its early days this company was kept busy exporting hay (of all things). Some years later following two or three very wet summers the Co-op's hay lorries were back in business again, this time importing hay from Counties Cork and Limerick, to help out it's suppliers in Counties Cavan and Monaghan.(25)

By 1977, milk intake had reached 14,000,000 Gallons and was still rising.(26)

In 1978, Bailieborough became involved in grain handling. For this purpose a Grain Store was set up in the Drogheda area. (27)

Milk intake in 1979 reached 22.9 million gallons, the same year saw the setting up of Emmets Cream Liqueur Plant. This was started in a premises originally built by the late

Tom Carroll as a fruit canning factory. Tom had initially set up in the old Market House.(28) I should point out that while operating in the Market House the fruit was all peeled by hand. Mc Elwaine's shop supplied the peelers. We always knew when Tom had changed staff, one time he would order 12 right handed peelers and 4 left handed peelers, next time it might be six rights and six left, or the next time again perhaps 8 left and 4 rights.

Back to Bailieborough Co-op.1980 was another busy year. The store in Henry Street was unable to cope with the volume of trade so in 1980 the Co-op bought the old Market House from Tom Carroll and set up their furniture, floor covering and fancy goods departments there. (29)

On the milk front, national figures for Milk production fell by 2.6%, but in Bailieborough Co-op intake increased by 11.4%.(30)

This was also the year that Bailie Foods was set up. Initially this was set up in 1965 to part dry the skim milk. Now, plans were changed and skim was dried right down to powder. This powder could either be used in animal feed or it could be exported to Third World Countries, to be reconstituted and used for human consumption.(31)

Another major advance in 1980 was the purchase of a 50% stake in Owens Dairies. This gave the Co-op a foot in the door to the Dublin market. A major advertising campaign was launched in the national press in an effort to catch a larger share of this market.(32)

In 1981, we saw the merger/takeover of NEF Co-op (North Eastern Farmers. This Co-op came out of the original Mullagh Co-op. They operated Retail Stores in Mullagh and Navan and also had a feed compounding mill in Mullagh. At harvest time NEF Co-op would have been one of the largest grain handlers in the area.

A large number of farmers would have been shareholders in both societies. I should mention that almost 20 years previously in the early 1960's a similar amalgamation had been mooted, but it fell through.(33)

The milk price in 1983 was 75p per gallon. Also in this year a "Technology Transfer Deal" was negotiated with an engineering firm in Bombay – India.(34)

In 1984 the cheese plant was set up on the site of the old Kingscourt Pig Mart. This was intended to produce "Feta Cheese", for supply to the Middle East countries. Due to the volatile political situation in those countries this venture was doomed from the start. The factory closed down after a couple of years. Twenty years later the building was still sitting there like a White Elephant. (35)

1985, started off well, but finished up very badly. At the start of the year Emmets were awarded a Gold Medal for the best cream liqueur. Also, early in the year the "Irish Horse Co-op" was launched. Whilst this was the "brain child" of Paddy O'Brien it was not intended to be financed by Bailieborough Co-op. A separate share capital drive was launched for the project. The initial response from both the farming and business community was very good. However, events which were about to unfold caught this venture unawares.(36)

Towards the end of 1985 two more projects were launched by Bailieborough, which would have far reaching effects, had they succeeded. The first of these was a bid to purchase from Brady Brothers in Cootehill, their livestock marts in Cootehill, Cavan and Mohill.

The second project was the purchase of a wholesale company in Dundalk. This company had, amongst other things, a timber import licence. In view of the colossal amount of building work that has taken place all over Ireland in the last twenty years, this would have been a licence to print money (if you will pardon the pun).

*Item No's – 15-36– Murray 77-86

As I said neither of these projects got off the ground, due to unforeseen circumstances. The bubble had not burst, but it had certainly become "pear shaped". At the time I felt that Bailieborough Co-op was a victim of circumstances beyond its control. In the intervening years, I have not seen or heard anything to convince me to change my mind.

In 1986 the "Smelly Stuff" really hit the fan. Everyone, Board of Management, Suppliers and the Public were completely taken unawares. It was unbelievable that a society that
twelve or eighteen months earlier had been hailed as one of the brightest stars in Irish Co-operative Organisation Society (I.C.O.S) firmament could now be in such a mess.
Paddy O Brien, General Manager resigned and an interim manager was appointed by the banks.

Everyone thought that Bailieborough Co-op could be traded out if its difficulties, but it was not to be. The wolves were howling for blood and so the entire Co-op was put on the market. The different enterprises were sold off, the hardware stores, N.E.F, Emmets etc.
When it came to the main business, the milk side of things did not go so well. At first the only man to express an interest here, was Frank Flanagan, Killeshandra Co-op. With the benefit of hindsight, perhaps more thought should have been given to the Killeshandra package. At the time the opposite was the case, in fact a lot of shareholders/suppliers were insulted by the terms of the offer. So much so, that when Larry Goodman, Beef Baron, threw his hat into the ring, he was welcomed with open arms.

There followed a long and at times acrimonious campaign between Killeshandra and Larry Goodman, to win the votes of the Bailieborough Co-op Shareholders. Eventually, Larry Goodman emerged the winner. After almost ninety years as a farmer owned co-operative, Bailieborough was now part of a public company.(9)
Larry did not take long to settle into Bailieborough. Before long he was off on the acquisition trail again. Firstly, he bought Westmeath Farmers in Mullingar, followed shortly after by Lagan Co-op in County Donegal.(10)

In 1989, Goodman started to get ambitious. He talked about forming a North Eastern Federation with the other Co-operatives in the area. At first the other Co-ops thought there was some merit in his suggestion. However, on further examination they realised that Goodman's plans went a little further than they intended. It became evident that Goodman

International – Food Industries PLC was to be the driving force and the other Co-op's would be very small fish, dancing to Goodman's tune. They decided that they would have to mount a counter campaign.(11)

In 1990, Lakeland Dairies was formed by the amalgamation of Killeshandra and Lough Egish Co-op. It was to be a three way marriage, but at the last minute, town of Monaghan Co-op decided to pull out.(12) At times the bouts got dirty. However, at a mass gathering of farmers, held in the Cavan Equestrian Centre the vote was overwhelmingly in favour of remaining as a co-operative rather than join Goodman PLC.(13)

*Items 9-14. MacDonald 96

Shortly after this, Goodman decided that the milk industry was not the "picnic" that he thought it to be. He then set about off-loading. In 1994 what were originally Bailieborough Co-op, Westmeath Farmers and Lagan Valley was sold to Golden Vale Co-op.(14)

At first Golden Vale took things "softly, softly". They wanted to re-establish the co-operative ethos in the area. Sometimes the suppliers felt that Golden Vale North was not getting as good a deal as Golden Vale South. However, they could do very little about it. They could not even threaten to transfer to Lakeland. After all the milk wars of the 1980's the Government had put in place terms and conditions for transfer. Firstly, one could only transfer in April every year, and then you had to give a months notice of intent.

Let us fast forward to the year 2001. Suddenly, out of the blue, Golden Vale Co-op sold their assets in this region to Kerry Co-op. Five minutes later, or so it seemed Kerry Co-op sold to Lakeland Dairies!!!. As I said, perhaps we should have given more consideration to Frank Flanagan's package. Here we are eighteen years later, having gone all round the houses, back to where we nearly were in 1988!!!.

I have written at length about the Co-operative Movement. Let me conclude this chapter with two pieces of "Useless Information".

The first concerns M/S J & L Stores, Henry Street, Bailieborough, (formerly Bailieborough Co-op Store). These premises were sold by the co-op in 1988, but there are farmers around today (2006) who still say they're going to the Co-op Store.

Secondly, in County Donegal the farmers speak of going to the "Cope", not the Co-Op as in other places. Back in the late 1960's and early 1970's there was a man called Pat Gallagher, who was a great supporter of the Co-operative Movement. So much so that he became know as Pat "The Cope" Gallagher. When he decided to enter politics 30 years ago, his name appeared on the Ballot Sheet as "Pat The Cope Gallagher". The name stuck and today if he is being reported in the press, he still gets the title "Pat The Cope Gallagher".

Chapter 30

In a Nutshell

Right through this book I have been showing the difference in farming during the sixty years between 1946 and 2006. In this final chapter I will try to summarise what has gone before.

Bailieborough Town

Right through the forties and into the sixties there would have been 70-90 cows fed and milked in town every day. These would have been in two or threes, but there were a few herds of 10-15 cows.

There would have been 150-200 pigs being fed in town at any one time. I myself used to feed 10-12 at a time.

As for hens, they were all over the place. Today one would be hard pushed to find even a pigeon!

In the 1950's we had fourteen shops in the town, selling animal feed. Today, we have only 2, and ironically the 2 premises selling feed today did not exist until the late sixties. Of course the Fair Day has long gone, as has the Middle Market (Pigs only) the Pork Market (2nd Tuesday in the month) and the Turkey Market at Christmas. Milk Carters with their horse and carts, (later tractors and trailers) are no longer causing traffic jams on the Main Street.

The present day articulated lorries would each carry as much milk as twenty of the old time carters with horses and carts. They no longer stop to do their "shopping" on their way through.

Rural Electrification

Opinions differ as to the cause of the most dramatic change to the farming community. Some say it was the arrival of the tractor. In my opinion, without a shadow of a doubt it was the Rural Electrification in the late 1950's and early 1960's. Prior to this everything had to be done by hand. The cows had to be milked by hand. Then with Electricity we got

Milking Machines. Next we advanced to Milking Parlours. These gradually became more sophisticated till we arrived at today's situation where we have computerised feeding, automatic cluster removals and even Robot Milking.

Rural Electrification also meant we could have water pumped around to all the sheds resulting in more livestock being kept. It also enabled farmers to set up intensified pig and poultry units.

Milk Yields – In the 1940's the average milk yield per cow, in Ireland was 380 Gallons per year. By the 1950's it had increased to 390 Gallons, in the 1970's it had jumped to 550 Gallons. Today the average would be between 1000 and 1200 Gallons, and yields of 3,000 Gallons, while not very common, have been recorded.

Milk Price – In 1950 the price of milk was 1 Shilling per gallon (Approx 4 Cent) Today (2006) the price is 27 Cent per Litre, equal to approx €1.23 per gallon.

Butter Making – Up to the late 1950's a lot of farmers churned their own butter. (See reference Grahams Hotel). In the 1970's this went out of fashion and people switched to buying "Creamery Butter". Sometime in the 1990's there was a limited return to home churning (Novelty). However I can't remember when I last ate home made butter. (Known as Country Butter)

Cattle Breeds –Back in the 1940's and 1950's we had mostly Shorthorn, Hereford and Aberdeen Angus with a few Jersey and Kerry. Over the intervening years, various breeds have been imported. The most important of these were, on the dairy side the Friesian and later the Holstein, and on the beef side the Charollais, Simmental and Limousins.

Hay Making – During the 40's and 50's hay would have been the principal form of Winter Feed. In those days from the time the hay was cut till it was safely in the shed could take up to three months. Nowadays there is much less dependence on Hay. However for those who still make hay, the advance in machinery makes this a lot easier. Now from cut to shed takes a maximum of 1 week.

Silage – In the sixties, Farmers started to switch to making Grass Silage. At the time silage was still fairly labour intensive (although not as bad as hay). The main advantage of Silage over Hay was that Silage was less dependent on the weather. However in those days a lot of attention had to be paid to temperature of the grass in the pit at the time of filling. The Temperature had to be checked in the morning and depending on the temperature you either put more grass in, or you rolled the pit and left it till the next day. I cannot remember the correct temperature required, but if the temperature was too high the resulting silage would be highly nutritious but not very palatable, and if not high enough it would be highly palatable but not very nutritious (or should that be the other way around). It could take a week to get the silage pit filled and covered.

Nowadays, things are different; the faster one can get the grass in and covered the better. With modern machinery it is possible to start silage making early in the morning and have the pit ready for sealing by the early afternoon.

Grain - Sixty years ago there were a couple of acres of corn grown on every farm. The yield per acre would have been approximately 15 CWT (750kgs) per acre. The crop was cut in August and the Thresher came in October/November. The field would probably have been ploughed with a couple of Horses.

In 2006 the corn is mostly grown on specialist tillage farms, consisting of 50,70 or even 100 Acre farms. The ploughing, sowing etc would all have been done by tractor, and the yield would be 3 – 3.5 Tonne per acre. Cutting with a Combine Harvester could start on a Monday and the field could be ploughed and ready for the next crop by Friday/Saturday.

Potatoes – Back in the 1940's – 1950's there would have been one or two acres of potatoes grown on every farm. The field would have been ploughed with a pair of horses the F.Y.M would have been spread in the drills with a manure fork or grape. Planting the seed potatoes or dropping would have been done by hand. Harvesting was done with a Horse Drawn Digger and the potatoes were collected by hand.

Potatoes not required for immediate use were stored in a pit of clay.

Fast forward to 2006, Specialist Farms, Bigger Acreage, planted and harvested by machine. The potatoes are collected by machine and for all intents and purposes are "Untouched by Human Hands".

Tractors – In my youth, the most popular tractor was the little grey Ferguson 20. By 2006, we have massive 100 Horse Power Brutes. The driver of the Ferguson 20 was exposed to all the elements. It did not even have a Safety Roll Bar. On the other hand the modern version has a heated cab, radio installed and all the modern gadgetry. It is even possible for one man to drive the tractor he is sitting in while at the same time drive another by remote control. Don't get me wrong, I don't mean driving on the public road! However, with ploughing or similar tasks it is possible to have one tractor following another.

Silage Contractors - In the late sixties, silage contractors were beginning to appear. At the time, the cost of getting one acre of grass cut and drawn into the pit was £4 -£5. Fast forward to 2006 and that same acre will cost €120. If the price of diesel continues to rise at the rate that it has been doing in recent months, it may soon cost a lot more than €120.

Sample Prices –

Year	Product	Cost
1978	12kg Hydrosan Plus (Dairy Detergent)	£6.50 / €8.25
1987	12kg Hydrosan Plus (Dairy Detergent)	£8.75 / 11.36
2006	12kg Hydrosan Plus (Dairy Detergent)	€34.95
1981	100 – 5' Paling Posts	£68.0 / €86.0
2006	100 – 5' Paling Posts	€267.00
1981	1 Tonne 27% C.A. N. (Fertilizer)	£125/€158.75
2006	1 Tonne 27% C.A. N. (Fertilizer)	€226.00
1954	1 CWT (50kg) Sow and Weaner Meal	£1.77 / €2.40
2006	25kg Sow and Weaner Meal	€7.50
1981	150 Meters Coil 60mm Land Drain Pipe	£30.00/ €38.0
2006	150 Meters Coil 60mm Land Drain Pipe	€96.00

WILLS's
GOLD FLAKE
for all occasions

Mr. Andy Gregson
Thomas

8th March 1957

MICHAEL CLARKE,
General Grocer,
Provision & Hardware Merchant,
Lower Main Street,
BAILIEBORO'.

'Phone: Bailieboro' 23.   Funeral Undertaker.

					21	12	0½
Mar	24	6 Bags Grass			11	5	0
May	4	1 Bag Stores 20/-	47/-	1	12	7	
	18	Bal. a/c 23 Bags		1	17	6	
Jun	22	3 Cw. Maize 7-6 /1x10 St. Meal		2	11	0	
		1 Cw. Calf Meal 5/8x8 oz cwt		3	1	6	
July	7	5/7 Cw. Super 5/9 Cw.		1	14	7½	
Aug	6	3 Cwt. Maize 1 Cw. Flour	357	1	4	6	
		2 Cw. Feedall 1 Cw. Pollar		2	9	6	
Oct	10	1 St. Bran			3	7	
Jan'53	28	1 Cwt. Meal			8	6	
		1 Cwt. Maize		1	10	6	
Mar	14	1 Cwt. Pure Maize		1	16	6	
	29	1 Cwt. Meal			8	6	
'1957	5	By Cheque	9 - 0 -	63	10	4	
Apr	5		15 - 0 -	24	10	0	

J & L STORES LTD.
Henry St.,
Bailieborough,
Co. Cavan

PHONE: 042 9665191 FAX: 042 9665033
EMAIL: SALES@JANDLSTORES.IE
WEB: WWW.JANDLSTORES.IE

Account No.:	
Salesperson:	0019802/T
	KAREN LYNCH
Deliver To:	

Docket No. CASH SALE
0019/005418
COLLECT
Date
Time   05/09/06
11:20:28

Account:
CASH DESK CASUAL CASH
J & L STORES LTD.
HENRY STREET
BAILIEBOROUGH
CO. CAVAN

PRODUCT	DESCRIPTION	UNIT	QTY	PRICE	VALUE
0093118	HYDROSAN PLUS 12KG	EA	1	34.95	34.95
0071001	27.5% CAN GRASSLAND 50KG	EA	20	11.25	225.00
0012006	LAND DRAIN 60MM 150 BLACK	EA	1	96.00	96.00
0090010	5" ROUND POST F	EA	100	2.67	267.00
0089302	LAKELAND SOW & WEANER 25KG	EA	1	7.50	7.50

Thank you for shopping at J&L Stores

Docket Total €
630.45

WEEE REG NO 008

RATE	GOODS	V.A.T.	VALUE
.00	232.50	.00	232.50
21.00	328.88	69.07	397.95
TOTAL	561.38	69.07	630.45

In conclusion, let me say - it never ceases to amaze me, when one considers all the good Agricultural land that has gone under concrete, for laying motorways, by-passes, domestic and industrial buildings, we can still manage to feed the nation and have some left to export as well.
WE REALLY ARE A GREAT LITTLE COUNTRY!

Bibliography

Items No 1-14	**Source**
Lakeland Dairies and the triumph of Co-operation 1896-1996	Author: Brian McDonald
Items No 15-36	
Bailieborough Co-op Group 1902-1977	Author: Aidan Murray B.A.g
Various Bailieborough Co-op Yearbooks 1978-1986	Author: Aidan Murray B.A.g